SpringerBriefs in Climate Studies

More information about this series at http://www.springer.com/series/11581

Marie-Jeanne S. Royer

Climate, Environment and Cree Observations

James Bay Territory, Canada

 Springer

Marie-Jeanne S. Royer
Geography and Earth Sciences (DGES)
Aberystwyth University
Aberystwyth, UK

ISSN 2213-784X   ISSN 2213-7858   (electronic)
SpringerBriefs in Climate Studies
ISBN 978-3-319-25179-0   ISBN 978-3-319-25181-3   (eBook)
DOI 10.1007/978-3-319-25181-3

Library of Congress Control Number: 2015953010

Springer Cham Heidelberg New York Dordrecht London

Printed on acid-free paper

Springer International Publishing AG Switzerland is part of Springer Science+Business Media
(www.springer.com)

Preface

The contents of this book are a reworking of my doctoral thesis which was completed in 2012. Writing this book presented an opportunity to translate and condense the results, making it more accessible to the communities who participated in the research.

The contents of this book explore changes that have been observed by the Cree and that appear in scientific data on the territory of the James Bay. The idea of combining these two types of data streams was the driving force behind the original project.

Acknowledgments

None of this research would have been possible without the enthusiastic members of the Cree Trappers Association and the Eeyou Itschee's community at large. More particularly I wish to thank Rick Cuciurean, Thomas Coon and Fred Tomatuk for their warm welcome and dedication to the subjects of traditional ecological knowledge and climate change. I also wish to thank Marc Girard for the great map which graces the inside of this book. And finally a big thank you to my friends and family who read and reread these chapters to polish them up.

Contents

Abbreviations

AR	Administrative region
CBD	Convention on Biological Diversity
CNG	Cree Nation Government
CO_2	Carbon dioxide
COP	Conference of Parties
CRA	Cree Regional Authority
CTA	Cree Trappers' Association
DGNQ	Direction général du Nouveau Québec
GCC	Grand Council of the Cree
GPS	Global Positioning System
IPCC	Intergovernmental Panel on Climate Change
IUCN	International Union for Conservation of Nature
JBNQA	James Bay and Northern Quebec Agreement
kV	Kilovolts
MEA	Millennium Ecosystem Assessment
NASA	National Aeronautics and Space Administration
NBR	Nottaway-Broadback-Rupert
NEQA	North Eastern Quebec Agreement
OURANOS	Consortium on Regional Climatology and Adaptation to Climate Change
SCBD	Secretariat on the Convention of Biological Diversity
TEK	Traditional Ecological Knowledge
UNEP	United Nations Environment Programme
UNESCO	United Nations Educational, Scientific and Cultural Organization
UNFCC	United Nations Framework Convention on Climate Change
WMGHGs	Well-mixed greenhouse gases
WMO	World Meteorological Organization

Chapter 1
Introduction

How do you start a book about climate change? The subject topic has grown rapidly, and it now feels that you can't turn around without hearing about it. It is broached in a myriad of ways from the purely scientific to the social, cultural, political, environmental, and the list goes on. In 2010, Anderegg et al. presented results indicating that 97 % of publishing climate scientists agreed with the Intergovernmental Panel on Climate Change (IPCC) that there was "unequivocal" warming of the climate system over the second half of the twentieth century and that "most" of this warming was "very likely" due to anthropogenic greenhouse gases (IPCC 2007; Anderegg et al. 2010). Such an interest in climate change is not surprising as increased climatic perturbations are having noticeable effects on our planet. However, for many climate change remains something that is distant both in time and place (Spence et al. 2012; Capstick and Pidgeon 2014).

This disinterest is partly because many of the severest predicted effects of climate change will only happen at a much later date, even though there are currently observable changes. It is easier for people that are living in the industrialised northern hemisphere to overlook some of the changes than people in developing regions as the latter are predicted to sustain harsher environmental and socioeconomic impacts (Alcamo et al. 2007; Cline 2007; Samson et al. 2011).

The object of this book is to help bridge this distance by identifying some of the changes that have already been observed. The James Bay territory in subarctic Canada, although slightly remote and off the beaten path, is part of an industrialised northern hemisphere country. The decision to work in subarctic eastern Canada was not a random one. The world's arctic and subarctic regions are at the forefront of change; with numerous studies showing that they are one of the first regions to encounter the impacts of climate change (Symon et al. 2005; Trenberth et al. 2007; IPCC 2013).

These impacts go much further than a change in temperature. Observed and projected changes include a decrease in sea ice, increased temperature, increased precipitations, warming of the permafrost and decreased snow cover. Close links have been observed between air temperature, precipitation, snow and ice

© The Author(s) 2016
M.-J.S. Royer, *Climate, Environment and Cree Observations*,
SpringerBriefs in Climate Studies, DOI 10.1007/978-3-319-25181-3_1

conditions, fauna and flora distribution, and many more ecological and environmental aspects, as well as with the indigenous communities in the Canadian and
circumpolar Arctic (Krupnik and Jolly 2002; Laidler et al. 2009; Ford 2009;
Weatherhead et al. 2010). In other words, changes are occurring on the structures
and functions of arctic and subarctic ecosystems and landscapes which in turn affect
land use, ecosystem services and ultimately the quality of life, the health and the
well-being of the concerned societies.

Aboriginal populations living in these regions are more at risk of feeling the
impacts of these changes on their way of life as they habitually still practice, at least
partially, traditional activities which are dependent on climate-sensitive resources
of biodiversity (e.g. hunting, fishing, harvesting plants and berries) (Riedlinger and
Berkes 2001; Furgal and Prowse 2008; Cavaliere 2009; Downing and Cuerrier
2011). These population groups are susceptible to important disturbances in their
supply chains, particularly those linked to their traditional subsistence activities. As
stated by Berkes and Jolly (2001) "Indirect effects of climate change, such as
changes to the availability of forage and water or the intensity of parasitic infec-
tions, may have an impact on [...] wildlife, and thus on the community harvest."
(p. 7) However even when quality and availability of a species has not changed, its
distribution, migratory patterns or habitats may have and in doing so rendered the
species inaccessible to populations that historically relied on them for subsistence.
An additional obstacle can occur when transformations in the local environment
places the species out of reach even when no direct changes have occurred to the
species itself, for example when modifications in environmental conditions such as
ice or snow increases the risks linked to the practice of traditional subsistence
activities. This can then lead to an increase in food insecurity, a loss of well-being
and community identity.

The direct and indirect impacts of climate change further exacerbate the
problems faced by an area where the local populations have experienced and are
continuing to experience important and rapid environmental and socioeconomic
transformations (Turner and Clifton 2009; Lemelin et al. 2010; Sayles and
Mulrennan 2010). Although these changes have been studied in the Arctic, the
subarctic has been the subject of much less research even though with its high
climatic variability it also faces substantial risks (Turner and Clifton 2009; Moser
2010). However transformations in the natural environment have previously been
identified as an important influence on the continuation of the traditional way of life
for aboriginal communities in the subarctic (Coon-come 1991; Masty 1991).
Additionally the study of the socioeconomic and sociocultural impacts of climate
change has had a relatively late start compared to the study of physical aspects of
climate change (Duerden 2004; Liverman 2008; Moser 2010; Wainwright 2010).
However because of the importance of the changes occurring in the subarctic, it is
clear that they will require the aboriginal communities to adapt their way of life to
the changing environment. It is with this in mind that we stress the importance of
understanding the changes taking place in the subarctic to allow for a well-adjusted
adaptation.

There are 614 indigenous communities in Canada, which accounts for 3.8 % of the Canadian population (Statistics Canada 2006). Indigenous populations are recording the largest population growth with an increase of 45 % from 1996 to 2006. For the same period the non-Indigenous population growth was 8 % (Statistics Canada 2008). The Cree First nation follows the same pattern with their population having almost tripled from 1971 to 2011 (Salisbury 1986; MAMROT 2011). In the eastern James Bay, the presence of the Cree First nation has been documented since at least the eighteenth century, where there is mention of them having divided the area into family hunting territory (Morantz 1978; Tanner 1983). This long occupation of the eastern James Bay by the Cree has allowed the development of traditional ecological knowledge (TEK) in relation to their traditional subsistence activities.

In addition to a significant population growth, the Cree First nation of the James Bay territory has experienced important changes in its environment. There have been a number of important land development projects in the area such as the James Bay hydroelectric megaproject. Beginning in the 1970s this ongoing project has led to the creation of eight hydro-electric stations and eight reservoirs and has a watershed that stretches over 176,000 km^2. Besides the obvious environmental changes, this project has led to many socio-economic changes inside the community, from the development of social and physical infrastructure, to a greater integration of the Cree into the southern economy.

For the members of the Cree First nation community of the James Bay territory (or Eeyou Istchee in Cree) traditional subsistence activities are intimately linked to personal identity and cultural values (Ohmagari and Berkes 1997; Tanner 2007; Power 2008). In 1978 the Cree Trappers' Association was created to protect the knowledge and culture associated with traditional hunting and trapping activities through knowledge sharing, as well as to develop ongoing research related to the environment and the monitoring of biophysical and anthropogenic changes in the environment. Most active Cree hunters and trappers belong to this association. They share their observations and knowledge of weather patterns and their fluctuations, and of environmental conditions. These observations are developed through sustained personal experience in the local area (Berkes et al. 2000; Sayles and Mulrennan 2010).

The case study presented in this book pulls from experiences and observations made by members of the Cree Trapper's Association. Through their continued use of the land based resources Cree hunters and trappers can offer insightful and timely knowledge of changes in their environment and of the weather patterns they encounter. Research in the Arctic and subarctic is moving towards a more participatory approach. It is moving away from simple documentation of the changes to the development of adequate strategies of adaptation and mitigation for a variety of fields from the ecological to the socio-cultural and economic. There is also a movement by numerous aboriginal communities to rightly request to be included in research undertaken on their lands, to be part of the research as equal partners. This is accomplished in part by taking into account their needs and traditional ecological knowledge (TEK). The integration of TEK and scientific knowledge in research allows a better identification of the interactions between climate change, northern communities and their environment (Symon et al. 2005; Gearheard et al. 2006; Pearce et al. 2009; Laidler et al. 2010). The combination of TEK and scientific knowledge is viewed by many researchers as the next step in participatory projects and it is with this thought in mind that this book explores the integration mechanisms needed to accomplish it.

The very large land surface area and low population density of the James Bay territory creates an impressive obstacle to systematic research (Meunier 2007; Turner and Clifton 2009; Brown 2010). They are part of the factors which have caused the James Bay territory to be plagued with a relative lack of long term climatic data when compared with other regions. Combining TEK and the available scientific data permits us to further develop and complete our knowledge of the region; creating a better understanding of the transformations caused by climate and a allowing us to delve deeper in the associated causes and consequences. In doing so it paves the way to better adapt mitigation and adaptation programs based on what has been observed locally and on the needs of the communities located in the area.

The information presented in this book is based on this approach. It documents the perceptions and observations of Cree hunters on climate change and on their environment. They were then combined with the available scientific data to look at what could be learned about how the environment and climate is changing in the James Bay territory. It then draws a multidisciplinary portrait of the forces at play in the region.

The book is divided into three parts. First the theoretical bases of climate change and TEK are established. Both these concepts have been surrounded by multiple debates, misunderstandings and questions. It is therefore our belief that any research on the subject must be approached with a clear understanding of where the author stands on the subject to avoid any potential confusion. Second the physical, social and historical aspects of the James Bay territory and the Cree that inhabit the region are explained. As change is a movement from one state to another and doesn't happen in a void, it only seems right to explain from where it started before explaining where it is going. The James Bay territory and the Cree have seen many changes both socioeconomically and environmentally in the last 60 years and they must be taken into account in any analysis of regional change. Third the case study itself and its results are presented. Last are the conclusions. As with many researches, this is not an endpoint but rather a step towards the future.

References

Alcamo J, Florke M, Marker M (2007) Future long-term changes in global water resources driven by socio-economic and climatic changes. Hydrol Sci J 52(2):247–227

Anderegg WRL, Prall JW, Harold J, Schneider SH (2010) Expert credibility in climate change. Proc Natl Acad Sci U S A 107(27):12107–12109

Berkes F, Jolly D (2001) Adapting to climate change: social-ecological resilience in a Canadian western Arctic community. Conserv Ecol 5(2):18

Berkes F, Colding J, Folke C (2000) Rediscovery of traditional ecological knowledge as adaptive management. Ecol Appl 10:1251–1262

Brown RD (2010) Analysis of snow cover variability and change in Québec, 1948–2005. Hydrol Process 24:1929–1954

Capstick SB, Pidgeon NF (2014) Public perception of cold weather events as evidence for and against climate change. Clim Change 122(4):695–708

Cavaliere C (2009) The effects of climate change on medicinal and aromatic plants. Herbal Gram 81:44–57

Cline WR (2007) Global warming and agriculture: impact estimates by country. Center for Global Development and the Peterson Institute for International Economic, USA

Coon-Come M (1991) Environmental development, indigenous peoples and governmental responsibility. Int Leg Pract 16:108–111

Downing A, Cuerrier A (2011) A synthesis of the impacts of climate change on the First Nations and Inuit of Canada. Indian J Tradit Knowl 10(1):57–70

Duerden F (2004) Translating climate change impacts at the community level. Arctic 57(2):204–212

Ford JD (2009) Vulnerability of Inuit food systems to food security as a consequence of climate change: a case study from Igloolik, Nunavut. Reg Environ Change 9(2):83–100

Furgal C, Prowse T (2008) Northern Canada. In: Lemmen DS, Warren FJ, Lacroix J, Bush E (eds) From impacts to adaptation: Canada in a changing climate 2007. Government of Canada, Ottawa

Gearheard S, Matumeak W, Angutikjuaq I, Maslanik J, Huntington HP, Levitt J, Kagak DM, Tigullaraq G, Barry RG (2006) "It's not that simple": a collaborative comparison of sea ice environments, their uses, observed changes, and adaptations in barrow, Alaska, USA, and Clyde River, Nunavut, Canada. Ambio 35:203–211

IPCC (2007) Summary for policymakers. In: Solomon S et al (eds) Climate change 2007: the physical science basis. Contribution of Working Group I to the fourth assessment report of the intergovernmental panel on climate change. Cambridge University Press, Cambridge

IPCC (2013) In: Stocker TF, Qin D, Plattner G-K, Tignor M, Allen SK, Boschung J, Nauels A, Xia Y, Bex V, Midgley PM (eds) Climate change 2013: the physical science basis. Contribution of Working Group I to the fifth assessment report of the Intergovernmental Panel on Climate Change. Cambridge University Press, Cambridge, p 1535

Krupnik I, Jolly D (eds) (2002) The earth is faster now. ARCUS, Fairbanks

Laidler GJ, Ford JD, Gough WA, Ikummaq T, Gagnon AS, Kowal S, Qrunnut K, Irngaut C (2009) Travelling and hunting in a changing Arctic: assessing Inuit vulnerability to sea ice change in Igloolik, Nunavut. Clim Chang 94:363–397

Laidler GJ, Elee P, Ikummaq T, Joamie E, Aporta C (2010) Mapping sea-ice knowledge, use, and change in Nunavut, Canada (Cape Dorset, Igloolik, Pangnirtung). In: Krupnik I, Aporta C, Gearheard S, Laidler GJ, Kielsen-Holm L (eds) SIKU: knowing our ice, documenting Inuit sea-ice knowledge and use. Springer, Dordrecht

Lemelin RH, Dowsley M, Walmark B, Siebel F, Bird L, Hunter G, Myles T, Mack M, Gull M, Kakekaspan M (2010) The Washaho First Nation at Fort Severn, The Weenusk First Nation at Peawanuck, Wabusk of the Omushkegouk: Cree-polar bear (*Ursus maritimus*) interactions in northern Ontario. Hum Ecol 38(6):803–815

Liverman DM (2008) Conventions of climate change: constructions of danger and the dispossession of the atmosphere. J Hist Geogr 35(2):279–296. doi:10.1016/j.jhg.2008.08.008

Masty D (1991) Traditional use of fish and other resources of the Great Whale River region. Northeast Indian Q 8(4):12–14

Meunier C (2007) Portrait and known environmental impacts of climate change on the James Bay territory. James Bay Advisory Committee on the Environment, Quebec

Ministere des Affaires Municipales, des Regions et de l'Occupation du Territoire du Quebec (MAMROT) (2011) Baie-James. www.mamrot.gouv.qc.ca

Morantz T (1978) The probability of family hunting territories in eighteenth century James Bay: old evidence newly presented. In: Cowan W (ed) Proceedings of the ninth algonquian conference. Carleton University, Ottawa

Moser SC (2010) Now more than ever: the need for more societally relevant research on vulnerability and adaptation to climate change. Appl Geogr 30:464–474

Ohmagari K, Berkes F (1997) Transmission of indigenous knowledge and bush skills among the Western James Bay Cree women of Subarctic Canada. Hum Ecol 25(2):197–222

Pearce TD, Ford JD, Laidler GT, Smit B, Duerden F, Andrachu M, Allarut M, Baryluk S, Daila A, Elee P, Goose A, Ikummaq T, Joamie E, Kataoyak F, Loring E, Meakin S, Nickels S, Shappa K, Shirley J, Wandel J (2009) Community collaboration and climate change research in the Canadian Arctic. Polar Res 28:10–27

Power EM (2008) Conceptualizing food security for aboriginal people in Canada. Can J Publ Health 99(2):95–97

Riedlinger D, Berkes F (2001) Contributions of traditional knowledge to understanding climate change in the Canadian Arctic. Polar Rec 37(203):315–328

Salisbury RF (1986) A homeland for the Cree: regional development in James Bay 1971–1981. McGill-Queen's University Press, Kingston

Samson J, Berteaux D, McGill BJ, Humphries MM (2011) Geographic disparities and moral hazards in the predicted impacts of climate change on human populations. Glob Ecol Biogeogr 20(4):532–544. doi:10.1111/j.1466-8238.2010.00632.x

Sayles JS, Mulrennan ME (2010) Securing a future: Cree hunters' resistance and flexibility to environmental changes, Wemindji, James Bay. Ecol Soc 15(4):22

Spence A, Poortinga W, Pidgeon NF (2012) The psychological distance of climate change. Risk Anal 32(6):957–972

Statistics Canada (2006) Population reporting an aboriginal identity, by mother tongue, provinces and territories. http://www40.statcan.ca/l01/cst01/demo38a-eng.htm

Statistics Canada. Canada Year Book overview (2008) Health: some health disparitiesand concerns. Minister of Industry, Ottawa. http://www41.statcan.ca/2008/10000/ceb10000_000-eng.htm

Symon C, Arris L, Heal B (eds) (2005) Arctic climate impact assessment. Cambridge University Press, Cambridge

Tanner A (1983) Algonquin land tenure and State structures in The North. Can J Nativ Stud 3(2):11–320

Tanner A (2007) The nature of Quebec Cree animist practices and beliefs. In: Laugrand FB, Oosten JG (eds) La nature des esprits dans les cosmologies autochtones. Les Presses de l'Université de Laval, Québec

Trenberth KE, Jones PD, Ambenje R, Bojariu R, Easterling D, Klein Tank A, Parker D, Rahimzaeh F, Renwick JA, Rusticci M, Soden B, Zhai P (2007) In: Solomon S, Qin D, Manning M, Hen Z, Marquis M, Avery KB, Tognor M, Miller HL (eds) Climate change 2007: the physical science basis. Contribution of Working Group 1 to the fourth assessment report of the Intergovernmental Panel on Climate Change. Cambridge University Press, Cambridge

Turner H, Clifton H (2009) It's so different today: climate change and Indigenous lifeways in British Columbia, Canada. Glob Environ Chang 19:180–190

Wainwright J (2010) Climate change, capitalism, and the challenge of transdisciplinarity. Ann Assoc Am Geogr 100(4):983–991

Weatherhead E, Gearheard S, Barry RG (2010) Changes in weather persistence: insights from Inuit knowledge. Glob Environ Chang 20:523–528

Chapter 2
Climate Change and Traditional Ecological Knowledge

2.1 Why Climate Change?

On June 23rd 1988, Dr James Hansen of NASA is called in front of the American senate's committee on energy and natural resources where he declares: "Global warming has reached a level such that we can ascribe with a high degree of confidence a cause and effect relationship between the greenhouse effect and observed warming." (Hansen 1988, p. 40) That is not to say that the basic science surrounding climate change is relatively new or that it started with Dr Hansen. The link between carbon dioxide (CO_2) and global warming has been established since the start of the nineteenth century (Fourier 1888; MacCraken and Luther 1985; Fleming 1998). However it launches a media storm and is picked up by numerous newspapers not least of which is the New York Times (Shabecoff 1988). The follow up to this media storm is that climate change has become one of the great questions of the twentieth and twenty-first centuries. Climate change has influenced political and social development since the 1980s on and even more so since the beginning of the twenty-first century. Nonetheless this increased prominence of climate science on the political scene has also had the effect of creating a politically charged debate surrounding climate research (Wainwright 2010; Liverman 2008). The science is ever more criticised and put under a microscope even while our understanding of the different effects and processes linked to global warming has been expanding and refined since the 1980s.

Dr Hansen was not alone in his concerns; the same year as his declaration the United Nations Environment Program (UNEP) and the World Meteorological Organization (WMO) created the Intergovernmental Panel on Climate Change (IPCC) with its stated role:

> [...] to assess on a comprehensive, objective, open and transparent basis the scientific, technical and socio-economic information relevant to understanding the scientific basis of risk of human-induced climate change, its potential impacts and options for adaptation and mitigation. (IPCC 2013a, p. 1)

© The Author(s) 2016
M.-J.S. Royer, *Climate, Environment and Cree Observations*,
SpringerBriefs in Climate Studies, DOI 10.1007/978-3-319-25181-3_2

Two years after its founding, the IPCC published its 1990 report, where it stated that the increase in the quantity of atmospheric greenhouse gases brought about by human activities were increasing the Earth's global temperature (Houghton et al. 1990). In its 2001 report the IPCC went even further, stating that the majority of observed climate changes during the last 50 years are attributable to human activity (Weaver 2004). The latest publication is ever stronger in its language: "It is unequivocal that anthropogenic increases in the well-mixed greenhouse gases (WMGHGs) have substantially enhanced the greenhouse effect, and the resulting forcing continues to increase" (IPCC 2013b, p. 661). The IPCC doesn't itself conduct any research, it only 'reviews and assesses' work done by climate scientists. Its publications and results are as a result often seen as conservative in their estimates to give them a better chance of being agreed upon by all reviewing members. Since its foundation the IPCC has produced five assessment reports with the latest having come out in 2014. Each subsequent publication refines the results of the previous one. It is therefore no surprise that the majority of the scientific community today agrees with the IPCC's position that human activity has had and is having an undeniable impact on observed climate changes (Hansen et al. 2000; NASCSCC 2001; Oreskes 2004; IPCC 2007a, b Anderegg et al. 2010).

However the IPCC is not the only organization concerned with climate change. There have been numerous reports published by national governments or their respective environmental agencies presenting the potential effects of climate change on their territories as well as on the global climate. In Canada, a project was put in place to link research centres working on climate change impacts in the Canadian Arctic and subarctic, which resulted in the creation of the ArcticNet network. There have also been numerous national and international meetings, conferences, and symposiums such as the Montreal and the Kyoto Protocols and of course the UN Climate summits. Climate change is now an inescapable presence not only in science research but in many aspects of the social sphere. Even the industry sector is slowly coming around to taking into account their impact on the climate (Anshelm and Hansson 2011).

It is hardly possible to discussion climate change without mentioning the media. The study of the mediatisation of climate change has become a field of its own. It has pervaded not only the news media but also the entertainment sector with documentaries (e.g. An Inconvenient Truth) doomsday films (e.g. Waterworld, The Day after Tomorrow) and even comedies (e.g. The Kingsman).

2.2   Why Is There All This Interest in Climate Change?

As sceptics often repeat, the climate has been changing since the world has existed. Why has climate change and more specifically anthropogenic climate change become such a key issue for scientists? Contrary to changes in climate attributable to natural causes (e.g. glacial periods) which happened over millennia with the exception of a few sudden events (e.g. 8,2 Ka Bp), the current observed climate changes are happening at an accelerated rate (Kobashi et al. 2000). This is where the cause for concern lies. Whereas natural climate change with its (usually) slower pace allows terrestrial life the opportunity to adapt itself to the new environmental conditions; the speed at which the current changes are occurring is taxing the biosphere's adaptive capacity. It is thus having an undeniable impact on vegetation, animal species (e.g. important loss of biodiversity, increase in pathogens) and on humans (Weaver 2004; Hansen 2005). As is often repeated by climate scientists, we aren't trying to save the Earth, the Earth will survive. What we are trying to save is the way of life and the ecosystems that we have grown accustomed to. In other words, we are trying to save ourselves and our quality of life (IPCC 2014).

With climate change transcending any and all borders it becomes necessary to develop concerted efforts for mitigation and adaptation measures, and a combination of broad global and targeted regional focused research. This cooperation although under the scientific banner is also a highly political one. There are many ways in which climate change is political. One such aspect is the melting of the polar ice caps which have far reaching political implications that go beyond the rising of sea levels as countries rush to identify who owns the North Pole (Gupta 2009; Murray 2012). Why the sudden interest in a cold piece of ocean? An economic interest of course, with the prospect of an ice free Arctic ocean many are seeing it as an alternate shipping route which will cut back drastically on time and money as well as holding the potential to access important quantities of oil. Many also argue that the consequences of climate change can include an increase in civil unrest and even war as communities and individuals fight over access to dwindling natural resources (Raleigh and Urdal 2007; Bernauer et al. 2012). Climate change is expected to increase the disparity gap between the global north and global south regions of the world (Rosenweigh and Parry 1994; O'Brien et al. 2006; Alcamo et al. 2007; Cline 2007; Samson et al. 2011). The cause for this is twofold; with the global north potentially getting benefits from the changes and the global south having less financial resources to deal with the negative impacts. Regions in the global north although projected to see some important disruptions from climate change, such as flooding, increased fire and extreme weather events, are also projected to incur beneficial changes such as longer growing seasons, and easier access to mining and energy sources. As the global north is comprised of highly developed regions with easy access to capital, they will in general be more easily capable of quickly modifying their production techniques, and their development to benefit from the new conditions while minimizing the undesirable effects. Regions in the global south however with no easy access to capital will end up feeling the effects of climate change much more harshly. Additionally some of the regions that will be affected are already under much environmental strain and close to their ecological tipping points making them vulnerable regions. As climate change is anticipated to increase some of the problems they are already facing, such as decreased precipitations in already arid regions, this could have very dire consequences indeed.

2.3   What Exactly Is Climate Change?

The IPCC's definition of climate change is "any change in climate over time, whether due to natural variability or as a result of human activity" (IPCC 2007b, p. 2). This differs slightly from the United Nations' Framework Convention on Climate Change (UNFCCC) definition which was adopted in Rio de Janeiro in 1992 where climate change is identified as "a change of climate which is attributed directly or indirectly to human activity that alters the composition of the global atmosphere and which is in addition to natural climate variability observed over comparable time periods" (UN 1992, p. 7). Although both definitions have merit, it is beyond the scope and the objective of this book to identify which observed changes can be attributable to man and which is caused by natural variability. The following chapters will therefore make use of the IPCC's definition.

The IPCC estimates that between 1951 and 2012 the Earth's temperature has risen by an average of 0.72 °C, with the associated effects and consequences of this rise in temperature varying based on the region of the globe and the season (Trenberth et al. 2007; IPCC 2013b). It is estimated that the changes in climate will continue to affect the global average temperature and precipitations, bringing with it an increase in the frequency and severity of extreme climatological events (Lehner et al. 2006; Alexander et al. 2006; Christensen et al. 2007). Climate research studying its evolution during the twentieth century have noticed an increase in number, severity and duration of droughts, severe storms (e.g. cyclones), floods, and forest fires (Davidson et al. 2003; Trenberth et al. 2007).

High latitude regions such as the Arctic and subarctic have sustained amongst the highest and most direct effects of climate change, with this trend projected to carry on in the near and medium future (Melillo et al. 1993; Payette et al. 2001; Berkes and Jolly 2001; Nichols et al. 2004). Among the consequences of rising global temperatures, estimates indicate that permafrost temperatures have increased in most regions since the early 1980s, that snow cover has decreases especially in spring, and that total precipitations have increased (IPCC 2007b; Mekis and Vincent 2011). Climate model projections see a continuation of these trends, with projected decreases in near-surface (upper 3.5 m) permafrost extent by 37–81 %, in snow cover by 7–25 %, and in sea ice cover for the Arctic Ocean as to be nearly ice free in the summer (according to certain scenarios) by the end of the twenty-first century (Christensen et al. 2007; Hansen et al. 2005; O'Neill and Oppenheimer 2004; IPCC science basis 2014). Lake and river ice cover periods are also projected to decrease by 30–120 days depending on the model (Fast and Berkes 1998).

Daily temperatures in the province of Quebec have risen in the south by 0.2–0.4 °C per decade since 1960; with this increase being more noticeable for minimum temperatures than for maximums (Zhang et al. 2000; Yagouti et al. 2008). This trend is maintained in future projections with a greater increase in the winter compared to summer temperatures. It estimated that by 2050, there will be an increase in temperature of 4.5–6.5 °C in the north of Quebec and an increase in total precipitation of 16.8–29.4 % in the winter and of 3.0 % and 12.1 in the summer (Desjarlais et al. 2010). Already, the increase in snow precipitation between 1957 and 2003 in the north of Quebec is identified as a cause of the acceleration in permafrost thaw (Payette et al. 2004; Vallée and Payette 2007). The opposite holds true for the south of Quebec where a decrease in precipitations is projected (Räisänen 2008; Brown 2010; Desjarlais et al. 2010).

Projections show that the centre of Quebec (which includes the territory covered by the boreal forest in Quebec and Labrador, where the Eastern James Bay territory is situated) should see an increase in temperature of 3.5–4.9 °C in the winter and of 1.8–3.0 °C in the summer, and an increase in precipitation of 12.0–22.9 % in the winter and of 1.1–6.9 % in the summer by 2050 (Desjarlais et al. 2010). Studies in the area have also shown significant tendencies towards an earlier date of ice retreat on the bay (Gagnon and Gough 2005). As the territory of the Eastern James Bay is a transition zone between climatic and environmental conditions of the south of Quebec and the Arctic, the flora, fauna and climatic conditions of the territory varies along a north-south gradient. Studies in the area have shown that an increase in temperatures combined with a decrease (or a too slight increase) of precipitation can have important effects on the region's fire activity (Flannigan et al. 1998; Girardin and Mudelsee 2008). "If increased fire activity in eastern Canada is a natural response to warmer climates than at present, then any trend toward higher temperatures not compensated for by significant precipitation increases in the future could lead to an increase in the fire risk over eastern boreal North America" (Hély et al. 2010, p. 6). A potential impact of an increase in the frequency of forest fires in the boreal forest in the Eastern James Bay would impact the ecosystem by accelerating the changes in vegetation cover composition (Flannigan et al. 1998).

One of the most talked about effects of climate change is the increase of the ocean water level from the polar ice caps melt. This increase is estimated to have already been happening at a rate of 3.1 mm per year for the years between 1993 and 2003 (IPCC 2007b). However in the case of the Eastern James Bay, after the retreat of the ice sheet covering it during the last ice age, the territory has been subject to an important glacial isostatic adjustment. Today, the rate of glacial isostatic adjustment for the region is still ongoing at a relatively rapid rate, estimated at 1.0 cm/year^{-1} and 1.3 cm/year^{-1} (Mitrovica et al. 2000; Lafortune et al. 2006; Lavoie et al. 2012). Therefore the coastal areas of the Eastern James Bay and Hudson Bay are less at risk of the anticipated effects of the rise in ocean water levels (e.g. increase in flooding risk and erosion) compared to other areas of the globe (Gough 1998).

Climate change can also influence the environment and biodiversity through events such as the extinction of animal and plant species and the modification of ecosystems (Mayer and Avis 1998).

> Projections of the impact of global change on biodiversity show a continuing and often accelerating species extinctions, loss of natural habitat, and changes in the distribution and abundance of species, species groups and biomes over the 21st century. (SCBD 2006, p. 71)

These impacts could directly influence the behaviour of migratory species (e.g. the Canada goose, the caribou) by modifying the climate conditions and ecosystem fauna, which in turn would affect the environmental cues used by the animals to identify the migration periods and rest areas (e.g. freeze/thaw of lakes and rivers, decrease in salt marshes). Other species (e.g. some insects, moose, and deer) might also modify their behaviour taking advantage of the warmer climate to extend their settlement area to the north (Berteaux et al. 2010; Desjarlais et al. 2010). The northern displacement of animals has the potential to increase the quantity of predators (e.g. wolf, coyote) in the region, putting these invasive species in direct conflict with the Nordic species (Herrmann et al. 2012).

2.4   Are These Projections Reliable?

The probable effects of climate change are numerous and difficult to predict. Studies that try to model or qualify the general characteristics of the globe often face problems linked to missing or low quantities of data. "The main reason is that in various parts of the globe there is a lack of homogeneous observational records with daily resolution covering multiple decades that are part of integrated digitised data sets" (Trenberth et al. 2007, p. 300). Even when the data is present they are not always homogenised (e.g. changes in the data collection standards, change in spatial distribution of stations) (Peterson et al. 1998; Mekis and Hogg 1999; GCOS 2003). It is therefore important and necessary to verify the quality and quantity of data on which these studies are based and to use the tools, methods, and calculations which limit potential errors (Vincent et al. 2002; Zhang et al. 2005; Alexander et al. 2006). Moreover, some bioclimatic prediction models do not always take into account factors other than climate changes which may influence species distribution (Davis et al. 1998a). This omission can create a bias in the results. It is preferable to use bioclimatic models which include the interactions between species and the biological dispersion of species (Woodward and Beerling 1997; Davis et al. 1998b).

The basic nature of prediction models is such that the results they present are difficult to validate in the short term. However, it remains essential to evaluate the methods used for these projections. This can be done notably by evaluating the precision of the method on past cases and the sensitivity of the analysis (Botkin et al. 2007). Some of these models therefore combine paleoecology with knowledge in climate change (Froyd and Willis 2008). Notwithstanding the major developments in the field of climatological modelling (Jones and Moberg 2003), climatic and bioclimatic projection models must be used with caution. When used correctly, the models and their projections can be very valuable to evaluate the potential amplitude of climatic changes and their effects on species distribution as well as to identify the species, habitats and regions the most at risk (Pearson and Dawson 2003). Even though some of these results are provided in this book, it is always important to keep in mind that these aren't presented as absolute certitudes but potential projections based on the information that is currently on hand.

2.5 Climate Change, Adaptation and Vulnerable Spaces

The vulnerability of a space is linked to a region's ability to sustain the current local way of life and to evolve in that context. It is often used in theories of economic development. The vulnerability of a space can be defined as "[...] the susceptibility of a coupled human-natural system to experience harm as a result of being exposed and sensitive to a perturbation (such as climatic hazard) and lacking sufficient response (coping and adaptive) capacity to deal with the perturbation (Moser 2010, p. 466)". A region is considered vulnerable based on a set of factors which creates an imbalance in the region's development. These factors could be specific to the natural or human environment (e.g. significant natural handicaps, insularity, transport network), of a structural nature (e.g. demography, industry type), of a cyclical nature (e.g. anything that creates an imbalance through the modification of the status quo), or of a social and cultural order (e.g. geographic mobility, basic level of training) (Gumuchian 1990). Additionally, the concept of a space's vulnerability is relative with the norms or averages used as references differing depending on the location and the time period. They change based on the region, the era or the social categories observed (Saucan 1999). Determining the vulnerability of a space therefore varies in time and space.

A high capacity to adapt is key to limiting the vulnerability of a space and increasing its resilience. This allows a society to continue to develop itself in the face of changes and even benefit from them. This is all the more important in the context of climate change where it is projected that the consequences of a rapid increase in temperature will have generalised harmful effects for the majority of the World's ecosystems (e.g. increase in invasive species, in pathogens and insects, permafrost thaw, changes in the fire regime). These in turn will have an incidence on human populations which depend on the affected ecosystems. The ramifications will be particularly noticeable in economies where subsistence activities are based on natural resources (Lemmen et al. 2008; Downing and Cuerrier 2011). Environmental changes are expected to amplify the loss of biodiversity, which combined with the changes in the ecosystem will cause an important increase in the management costs of future societies, especially in ecosystems where the resilience limit is exceeded (Fast and Berkes 1998; Chapin 2000). Adaptation to change requires a capacity to challenge the values and practices by the concerned communities.

The importance of adaptation used in combination with mitigation efforts to face the problems linked with climate change has been raised at many levels (Lemmen et al. 2008). Numerous agencies and programs have been created to study this question. The UNFCCC alone has instituted several programs such as the: National Adaptation Programmes of Action, the Adaptation Fund under the Kyoto Protocol, the Least Developed Countries Fund, and the Special Climate Change Fund (Ford et al. 2007). These programs join various initiatives at the national, provincial and territorial levels which are developed following the article 4.1.b of the UNFCCC and article 11 of the Kyoto Protocol where, among other points, the signing members agreed to facilitate adaptation to climate change. Adaptation is important as it limits the negative effects associated with climate change (Rothman et al. 1998). It is therefore necessary to not only mitigate climate change but also to adapt to the changes that we will not be able to avoid.

Indigenous people that live in the Arctic and subarctic regions traditionally depend on the natural resources found there (Furgal and Seguin 2006). The regions in elevated latitudes are particularly vulnerable to climate change and its consequences on the ecosystems, which subsequently has an impact on a community's vulnerability (O'Brien et al. 2004; Herrmann et al. 2010). The traditional way of life for many of these communities is based on hunting, trapping and fishing and thus on the resources of the ecosystem. A modification of the ecosystem can lead to changes in habits and customs and on the capacity of indigenous people to supply themselves in traditional foods (Downing and Cuerrier 2011). Animals, which are often the base of traditional diets in the Arctic and subarctic, are sensitive to the climate and a change in climatic conditions could influence their distribution, proliferation, their health as well as their migration patterns (Ford 2009). The populations that inhabit these regions often have to travel large distances over ground and ice. A change in the environmental conditions could by consequence create difficulties in accessing the resources and increase the risk associated with land and ice travel (Berkes and Jolly 2001; Tremblay et al. 2006). Already many researchers have shown a decrease in the time where lakes and rivers are covered in ice in many Arctic and subarctic regions of Canada (Furgal and Prowse 2008).

> On the one hand, environmental change is integral to the daily lives of northern peoples, and a capacity to change is part of their livelihood systems. On the other hand, extreme events and unusual fluctuations create safety hazards as well as adaptation problems, and little is known about their cultural, social and economic limits to adaptability. (Fast and Berkes 1998, 207–8)

Indigenous communities must therefore adjust their knowledge to adapt to the new singularities of environmental phenomenon (Gearheard et al. 2006). To obtain a complete picture of the situation, it is necessary to take into account all the factors which determine the situation of the Cree of the Eastern James Bay.

> Fully understanding the real importance and potential severity of climate change for any location (both the impacts experienced locally and those affecting other regions but impacting local communities and sectors indirectly) requires placing climate change into the real-world context of multiple stressors, on-the-ground vulnerabilities, and the actual capacity of communities, businesses, and local and state government institutions to respond to rapidly unfolding changes in the physical and social environment. (Moser 2010, 467)

The Eastern James Bay has seen numerous changes over the last 70 years. These have touched many aspects of Cree life from social change (e.g. social organisation, population growth, and housing) and technological advances (e.g. snowmobiles, helicopters, GPS) to environmental change (e.g. large scale hydroelectric projects, mining and road construction, climate change). This has placed the Cree in a context of perpetual change and increased the need for rapid adaptation.

2.6   What Is Traditional Ecological Knowledge?

The concept traditional ecological knowledge (TEK) as a tool for strategic human adaptation to the environment has interested many researchers since at least the 1950s (Toledo 1992; Herrmann et al. 2010). Today the first hurdle faced by researchers wanting to work with TEK is often what to call it. Over the years many names have been proposed to replace the term TEK including *indigenous knowledge, local knowledge, local indigenous knowledge, community knowledge, aboriginal traditional knowledge* and *aboriginal knowledge* (Crawford et al. 2010). The main criticism associated with the terminology is its inclusion of the word 'traditional'. This word is taken by many as having a negative connotation. They suggest that it implies that this type of knowledge is simple, static and unevolved. With TEK often being applied to knowledge systems held by indigenous populations some have complained that it devaluates the knowledge type (Berkes et al. 2000; Houde 2007). I would argue that this is a simplistic and erroneous linguistic association. In the context of TEK, 'traditional' is used to refer to a cultural continuity which is transmitted through generations by the values, beliefs, conventions and practices of a community. Without its inclusion we lose the qualifier which rests on the long term and continuous aspects of this knowledge type. I would further argue that the other proposed terms are often reductionists in that they limit themselves by the use of the words indigenous or aboriginal. While indigenous groups often possess detailed and extensive knowledge on the local environment the singling out of these communities is detrimental to these populations. It holds them up as different and feeds into the myth of their 'otherness' while at the same time excluding other types of populations who through long term contact with the environment also possess extensive knowledge on the local environment. It would therefore be counterproductive to linguistically and by designation limit the expectation of this type of knowledge as belonging only to aboriginal populations. Not only does it restrict the potential uses of TEK in research and adaptive management, it also risks further stereotyping indigenous/ aboriginal populations. The use of the terminology TEK is thus preferable as it doesn't explicitly refer to or single out one type of community.

While the debate has somewhat slowed with a majority of current literature having settled on the use of TEK as the term of choice, synonym terminology as listed above is still used. The preponderance of the use of the terminology TEK in the literature was aided by its inclusion in the works of many worldwide organisms such as: the now defunct Traditional Ecological Knowledge Working Group of the International Union for Conservation of Nature (IUCN); the 1987s Brundtland Report: 'Our Common Future' from the United Nations Commission on Environment and Development; and the 1987 Convention on Biological Diversity multilateral treaty (WCED 1987; Berkes 1993; UNSG 1993; Bonny and Berkes 2008). For the remainder of the book we will continue to use the terminology TEK. Now that we know what to call it an important question remains: What is TEK?

TEK is a large and encompassing concept open to a few interpretations; it can refer both to the local knowledge of indigenous populations as well as to non-indigenous ones, because at its core it rests on the development of knowledge through long-term occupation and extensive interaction with the physical environment (Berkes 1993; Huntington 2000; Usher 2000). Indigenous groups are often seen as the prime example as this type of knowledge holders as they have had large scale uninterrupted contact with their local environment. However groups that are not typically seen as indigenous have also been shown to have had uninterrupted contact with their local regions, one such example being herding and farming communities in Europe.

TEK is a complex system of knowledge, practices and representations held by a society on the environment and ecosystem that surrounds them (Nakashima and Roué 2002). This complex system develops itself over a long time period spent occupying the same territory. As described by Maffi (2001, p. 11):

> But the overall consensus is that small local groups with a history of continued and unchallenged occupation of given territories will over time tend to develop and maintain detailed and accurate knowledge about their ecological niches, as well as about sustainable ways of extracting and managing natural resources.

This quote points out well the localised and long term aspects of TEK. The local communities through their use and experiences learn to identify their environment, having detailed knowledge on various aspects such as the species which inhabit it, especially the ones beneficial and detrimental to their communities, the layout of the terrain, and the weather. As these societies rely extensively on the local environment to sustain themselves their long term presence in the area often points to them having developed ways of interacting with the environment which allows them to extract resources in a way that is sustainable. Note that this equilibrium is not absolute. Changes in technology, climate, environment and/or knowledge can unbalance it making it necessary for the local communities to readapt while taking these changes into account.

Spirituality is often strongly linked with TEK knowledge as it plays a key role in managing societal interactions with the ecosystem (Lyver et al. 2009; Singh-Negi 2010). It assists in the creation and definition of strict and complex frameworks of practices and rituals; which in turn limits the use and access to resources (Scott 1989; Western et al. 1994). In doing so, spirituality and its related organisations help manage the conservation of available resources. Therefore as explained by Berkes (2008) TEK can be seen as the combination of:

- Knowledge (e.g. specie identification, taxonomy, life stories, distribution, behaviours, etc.);
- Practices (e.g. tool development, techniques, use of resources for commercial, medical, religious, or nutritional reasons, etc.);
- Spiritual beliefs (e.g. values system, ethics, symbolic meanings, perceptions, etc.).

These three elements are nested in one another and are strongly interrelated with each one aspect reinforcing the other.

A vital aspect of TEK, which cannot be stressed enough, is that similarly to other knowledge types it is constantly evolving. The knowledge contained in TEK is fluid, it develops and grows while being transmitted from one generation to the next (Berkes et al. 2000; Berkes 2008; Watson et al. 2008). It is normal and even desirable for each generation to reinterpret passed down knowledge so that they are better equipped to face new challenges and seize new opportunities (UNESCO 2009a). This capacity is even more important in a world facing rapid environmental and technical changes.

This knowledge does not develop in a vacuum. Communities possessing TEK are in contact with each other and with scientific communities. Not only does TEK evolve using knowledge gained independently by each new generation; it also integrates knowledge obtained through exchanges with other TEK communities and the external world. Most of the studied traditional knowledge communities have been in contact with scientific knowledge since at least the fifteenth century (Agrawal 1995). This knowledge sharing is one of the keys to a community's survival. It allows them to integrate new information, but it is not integrated wholesale; it is adapted and reshaped to better include it into the receiving community's TEK system. Berkes (2008, p. 7) sums it up perfectly when he defines TEK as:

> [...] knowledge, practice, and belief, evolving by adaptive processes and handed down through generations by cultural transmission, about the relationship of living beings (including humans) with one another and with their environment.

It is with this vision and this definition of TEK in mind that the research was undertaken on the relation between TEK and the biodiversity and on how to combine it with scientific research to reach a new way of understanding and studying the ever changing environment.

2.7   TEK and Biodiversity

TEK and the different interactions of communities with their surrounding environment are increasingly perceived as an important source of information on local biodiversity (Turner et al. 2000; UNESCO 2003; Greene 2004). Humans have been consciously and unconsciously shaping the environment around them for millennia, blurring the limits between nature and culture. Examples of human activities which have modified biological diversity are plentiful (Mitchell and Brown 2002). One such example is the numerous species and varieties of plants which have been developed by the different societies to satisfy social and environmental needs (Daily 1997; Gari 1999; Pinton 2003; Berkes and Turner 2006).

The vision of including human beings as an integral part of the natural environments is slowly overtaking the one that sees them as obstacles in the rational and efficient use of an environment's resources (Agrawal and Gibson 1999; Posey 2002). It has gained traction among many researchers and international organisations including the United Nations (Colchester 2004; Berkes and Turner 2006). In this vision, TEK and biodiversity are linked in a dynamic interrelation.

Local communities, through their long term settlement in a particular area have developed an intimate knowledge of the local environment which they know how to exploit in a variety of ways; they use the various species of fauna and flora available to answer many of their basic needs (e.g. dietary, pharmacological and clothing) (Kumar et al. 2004; Descola 2005). TEK is therefore part of the decision making process of their daily lives (UNESCO 2009b). Each culture possess an interrelated group of representations, knowledge, rules, and cultural and spiritual practices linked to nature and the environment which is unique to it and has been developed through long term and sustained contact with the surrounding environment and biodiversity (Posey 1999). In this situation a culture's spiritual and cultural aspects play a role in biodiversity protection (Maffi and Woodley 2010).

It should be noted that some researchers question this vision and are sceptical of the importance of TEK in the protection of biodiversity. Their argument is that although these groups possess a specialised knowledge of their environment, they do not always possess a way of life that is beneficial to it (Diamond 1993; Redford and Stearman 1993; Chapin 2000). As presented with the case of the Moose River Basin in Canada by George et al. (1995, p. 85):

> [...] transformations of cultural tradition [...] the shift toward efficient killing of large numbers of waterfowl, and quick, short (by snowmobile, even aircraft) sorties for large game, sustains a high percentage of bush food in the diet but changes the strategic and ethical context of hunting.

We would be remiss if we didn't acknowledge that the technological advances witnessed by the indigenous populations of northern Canada in the 1960s and 1970s (e.g. snowmobiles, radars, motor boats, helicopters, refrigeration) have caused changes in the practice of traditional subsistence activities (Tsuji and Nieboer 1999; George and Preston 1987). They have among other things diminished the time needed to spend hunting by increasing travel and hunting efficiency and increasing the number of kills. Although it is important to avoid falling for the myth of a perfect relation with nature, it remains that local populations, through their long term experience on the land have developed an in-depth knowledge of their environment (Alcorn 1993; Toledo 2001; Nadasdy 2005; Folke et al. 2007).

TEK and language are often interconnected, and many authors have established a correlation between areas with rich and developed biological diversity and linguistic diversity (Mülhäusler 1995; Sutherland 2003; Maffi 2005). Language is therefore considered by many as a strategic indicator between biological diversity and cultural diversity (Maffi 2005; UNESCO 2003; Butchart et al. 2010).

As explained previously, a change in one or many conditions, be it in the natural environment (e.g. increased rainfall, decreased presence of a particular species) or in the human environment (e.g. important increase in population size, modernisation of the way of life) can unbalance the Human-environment relation. The

Millennium Ecosystem Assessment (MEA) is one of the largest evaluations of the relationship between human well-being and ecosystems and included the participation of 1,300 experts from 95 countries, took place between 2001 and 2005. It evaluated 24 ecosystem services and bears witness to the imbalance of many of them (SCBD 2006). Of the 24 ecosystem services 15 services, including drinking water supply and pollination, are degraded (MEA 2005). The degradation of these services and the loss of biodiversity have an impact on the quality of life of the affected populations. A loss in biodiversity upsets the ecosystem and renders it more vulnerable to shocks and disturbances. The affected ecosystem is then less able to provide the services that humans need (SCBD 2006). Societies which depend quite heavily and directly on their environment and the services provided by the ecosystems, such as the Cree Nation of the Eastern James Bay, are among the first to feel the effects of this degradation and imbalance.

However, limiting our vision of nature based only on the services it offers reduces nature to a material commodity (Braun and Castree 1998; Reveret et al. 2008). Because of the nature-culture interrelation, the effects felt when a culturally significant or emblematic species disappears go beyond the obvious material consequences to touch the cultural domain; it can have repercussions on the language, the traditions and the intergenerational communication. It can even affect the very foundation of a society (Braun and Castree 1998; Berkes 2008).

The concept of nature is not independent of humans but rather a human creation that varies based on time and space (Cronon 1996). This vision of nature is increasingly being put forward in geography. As stated by Bakker and Bridge (2006, p. 6):

> In cultural geography, for example, 'rethinking the object' is invoked as a means of reincorporating (and rediscovering) the materiality of everyday life, or of accessing the relationships between commodities, cultural identity and the politics of value.

The relationship between communities and biodiversity is a two-way street (White 1995). As much as a community can influence the survival and proliferation of certain species, human institutions (be they political, religious, social or economic) are too influenced by the available natural resources provided by the biological diversity of the local ecosystem (UNESCO 2003).

With the numerous changes that face today's societies, it is important to study the impacts that these changes have on the social organisations:

> A critical contemporary need is to understand how the construction of knowledge societies and the transmission of ethno-ecological knowledge are changing as people alter their lifestyles in the face of pervasive demographic shifts: migration to seek higher standards of living elsewhere, displacement from homelands under pressure from conservation initiatives and agricultural development, and flight from climate change and natural disasters. (Herrmann et al. 2010, p. 102)

A large part of TEK is created by direct contact with nature through work; these changes can even influence the concept of nature (White 1996).

The Convention on Biological Diversity (CBD) recognises the potential uses of TEK and traditional methods for the conservation and sustainable use of biodiversity (SCBD 2006). While at the same time social diversity and TEK are identified as being a part of biodiversity on the same level as genes, species and ecosystems (Aubertin 2002). The Convention on Biological Diversity at Rio de Janeiro in 1992 takes a strong position for the protection of traditional knowledge in its article 8 (j) which states:

> Each contracting Party shall, as far as possible and as appropriate: subject to national legislation, respect, preserve and maintain knowledge, innovations and practices of indigenous and local communities embodying traditional lifestyles relevant for the conservation and sustainable use of biological diversity and promote their wider application with the approval and involvement of the holders of such knowledge, innovations and practices and encourage the equitable sharing of the benefits arising from the utilization of such knowledge innovations and practices. (SCBD 2006, pp. 176–7)

To be able to achieve the objectives listed in this article, the signing countries must deepen their understanding of the traditional knowledge present on their territories. They must also develop institutional and legal instruments to allow them to conserve and protect this information source (Roussel 2005).

Even though there is consensus on the importance of biodiversity there remains numerous discussions on the elements that should be protected and the ways to achieve this protection. A particular element of biodiversity can have significant cultural value for a community and have a completely different level of importance for an ecologist or a biologist (Carolan 2007; Caillon and DeGeorges 2007). Each country has one or many indigenous groups that differ by size, language, culture, traditions and numerous other socio-economic factors (Ohl et al. 2010). TEK being linked to the ecosystem in which a group lives, these notions are dynamic and the processes that govern the Human-environment relationship vary greatly based on the epistemological framework of each society (Peloquin and Berkes 2009). There is therefore no absolute answer on: *what do we need to protect, how we need to protect it* and *for whom we should protect it* (Aubertin et al. 1998; Patenaude et al. 2002; O'Flaherty et al. 2008).

2.8 Combining TEK and Scientific Knowledge

Botanists were among the first researchers to show an interest in TEK during the eighteenth century. TEK remained their domain for most of the nineteenth and twentieth century. However, the 1990s saw a renewed and increased interest in TEK which expanded beyond local indigenous groups and the researchers working with those communities (Agrawal 1995). This renewal of interest helped spur on the creation of related international organisations such as the Traditional Ecological Knowledge Working Group of the IUCN and other large scale projects such as UNESCO's Man and the Biosphere (Posey 1999). These initiatives increased the research done on the subject and the dissemination of knowledge to reach the popular imaginary through various media such as the Time magazine cover titled Lost Tribes, Lost Knowledge (Linden 1991). In science, TEK is no longer exclusively relegated to anthropology, sociology or geography. Instead it is integrated into numerous research disciplines such as ecology, medicine, botany, agronomy, computer science, landscape planning and management (Warren et al. 1993). At the same time, the different actors in biodiversity conservation realised that to be successful, an approach must be multidisciplinary and include the participation of local communities and the traditional knowledge (Nakashima 1993; Fraser et al. 2006; Nevins et al. 2009).

As we have already explained, TEK is created by integrating multiple information sources. However the advantage conferred by this combination is not unidirectional. Scientific knowledge can also obtain positive results by incorporating traditional knowledge possessed by communities (SCBD 2006; Stephenson and Moller 2009). The integration of TEK and scientific knowledge is particularly useful in isolated regions of the globe where scientific research is sporadic and limited. In these areas, TEK can be used to complete and enrich scientific knowledge of the regions (Moller et al. 2004; Lyver et al. 2009).

TEK is particularly useful in the identification of species and fauna and flora varieties, in animal behaviour, and in the life cycle of ecosystems (Alcorn 1989). This knowledge can then benefit society through agriculture, habitat restauration, the study of environmental impacts of development projects and in resource management (Menzies 2006). In North America, contrary to popular belief developed during the colonial period, many hunter-gatherer groups practice a complex management of their habitat using fire succession cycles, a rotation in resource exploitation as well as other tools and techniques (Cronon 1983; Turner 1999; Johnson 1999; Davidson-Hunt 2003). Their knowledge of vegetal and animal species present in the different stages can be employed for the management and conservation of ecosystems (Berkes and Davidson-Hunt 2006). Taking into account the socio-cultural framework allows for an increased comprehension of the roles played by the local biodiversity and an increased appreciation of local traditions by the community (Bhagwat and Rutte 2006; Jacqmain et al. 2007; Waylen et al. 2010). Increasing the valorisation of local traditions strengthens a community's resilience. As stated by Davidson-Hunt and Berkes (2003, p. 17):

> The point, however, is that the loss of resilience is not only social, but rather social-ecological in nature, because of the coupling between people and the environment in which they live.

As many authors have already stated, the destruction of ecosystem spaces often starts with a breakdown of the traditional management processes (Aubertin et al. 1998). These realisations have helped to increase the acceptance of co-management of resources, through a better understanding of the benefits of integrating TEK into scientific knowledge (Fraser et al. 2006; Plummer and Armitage 2007; Jacqmain et al. 2012).

In Canada one of the greatest leaps for the acknowledgement of the importance of TEK came in the 1970s when the judge Berger held public consultations for the pipelines and dam projects in the Mackenzie River valley. He waded through almost a thousand testimonies in seven different languages from inhabitants of northern Canada. In 1977 these public consultations led to the putting on hold of the project for 10 years to allow the Indigenous populations to negotiate an agreement (Roué and Nakashima 2002). They brought to the forefront many questions on the impacts of the construction of pipelines in arctic and subarctic regions of Canada and on the rights of the region's indigenous populations. They helped start a movement that places TEK as complementary to scientific knowledge.

This has raised the question as to how to classify TEK. Many researchers want to group TEK on the same level as scientific knowledge under a general label of knowledge. Others think that it is necessary to stress its specificity to help distinguish them; TEK being, as per its nature, subjective whereas scientific knowledge, as per its nature, looks to attain objectivity (Usher 2000; Turner and Clifton 2009; Stephenson and Moller 2009; Dickison 2009). TEK is strongly linked to the cultural identity of each of the communities it comes from. It is not only a type of knowledge it is also a way of life and a way of thinking. Science knowledge on the other hand is generally perceived as the knowledge provided by modern science which transcends the local context and looks for universal rules (Van Eijck and Roth 2007; Berkes 2009).

Put simply, the scientific method looks to identify general explanations of cause-effect relations in the form of a hypothesis. These then generate specific statements which once checked either confirm or infirm the starting hypothesis (Crawford et al. 2010). However many are debating this objectivity. Haraway in her book, 'Primate Visions', states that: "any scientific statement about the world depends intimately upon language, upon metaphor" (Haraway 1989, p. 4). Even though science tends towards objectivity, it is not possible for it to completely extricate itself from its environment as the epistemological framework of science cannot be separated from its socio-political framework (Rousse 1987; Latour 1991; White 1995; Forsyth 2003; Budds 2009). As such Latour indicates that western science has developed a separation between pure knowledge and knowledge linked to society but that this is an artificial separation as all science is linked to the society which holds it (Latour 1987). Others go much further accusing science of representing a masculine (some go so far as saying sexist) and westernised vision of the world (Martin 1991).

The concept of TEK by its description is subjective in nature however this subjectivity doesn't imply that it is wrong. Among arguments for its value, it can be noted that TEK is based on observations made in the field, an approach also used in modern sciences (Agrawal 2009). During the legal procedures for the development of the James Bay, when a Cree witness was asked to swear to tell the truth, his translator stated: "He does not know whether he can tell the truth. He can tell only what he knows" (Richardson 1976). This statement shows an important concept in Cree culture which puts faith in what has directly been observed rather than in what has been read or said by others. Truth is not something external that exists in itself but rather something that is developed through observations and teachings (Preston 2002; Peloquin and Berkes 2009).

The debate on where TEK fits as a knowledge system is best explained by Stephenson and Moller who have stated:

[...] it would be absurd to assert that all knowledge systems are equally powerful for all
uses and questions, but we hasten to add that asserting one system to always be more
powerful or legitimate than the other for all types of problems would be equally absurd.
(Stephenson and Moller 2009, p. 143)

In this way both TEK and scientific knowledge should be seen as complementary
but non identical forms of knowledge which can be used through multidisciplinary
approaches to both of their benefits.

There have also been debates on the pertinence of researchers external to a
community undertaking studies on that community's TEK, some arguing that TEK
is almost always taken out of context, badly interpreted or that its attributed
meaning is false or wrong in some way (Stevenson 1996). The common answer
to this type of criticisms is that although the information is interpreted, something
which is fundamental to research, and that this interpretation is done inside a
specific approach, the differences arrived at are mainly on the style rather than on
the substance (Wenzel 1999). The solution to this problem is to be well
documented, include consultation and cooperation between researchers and communities, and to follow a strict process to ensure the utmost faithful interpretation of
the data obtained (Calamia 1999; Usher 2000; Weatherhead et al. 2010). Most
importantly the objective must not be to modify the belief or merit systems of either
group (i.e. scientific community or local community) but rather to use the data
provided by both groups to attain a coherent and significant whole; as the goal for
any research is to provide a faithful interpretation of data linked to a specific
phenomenon (Higgins 2000; Cronin et al. 2004). It is with this in mind that the
case study presented here aims to show that by using both knowledge systems it is
possible to obtain more information on a given phenomenon.

TEK is increasingly being looked at by the scientific community to expand our
knowledge of how the communities living in arctic and subarctic regions are being
affected by climatic change (Weatherhead et al. 2010). Riedlinger and Berkes have
identified different areas where TEK can be integrated to scientific knowledge to
increase understanding of climate change in the Canadian Arctic and subarctic.
They state it can be used as:

[...] a source of climate history and baseline data; in formulating research questions and
hypotheses; as insight into impacts and adaptation in Arctic communities; and for long-
term, community based monitoring. (Riedlinger and Berkes 2001, p. 315)

Other researchers have agreed with them stating that TEK can be used to obtain
data linked to the environment, to climate, to species and to habitats, as well as to
contribute to the development of climate change adaptation models (Duerden 2004;
Turner and Clifton 2009). By combining TEK observations data to scientific data,
we can arrive at conclusions that allow us to better understand the current situation
and to develop new avenues of research, adaptation and mitigation methods and
better suited management programs (Freeman 1992; Tyrrell 2006; Pearce
et al. 2009; Weatherhead et al. 2010). Usually this combination of data is done
using qualitative data for TEK and quantitative or qualitative data from scientific
knowledge (Berkes et al. 2000; Tremblay et al. 2006; Laidler 2006). The TEK data

can be obtained using various techniques including interviews, questionnaires, joint field work and community workshops (Huntington 1998; Huntington 2000). This data is then triangulated with the scientific data to verify the information and to interpret it.

This is the approach that was taken in the following case study. TEK was used as a source of information on historical data about climate and the species it influences, on current data on the impacts of the changes to subarctic communities, and as a way to develop future research questions. This research aimed to further the understanding in the ways to combine TEK and scientific knowledge.

References

Agrawal A (1995) Dismantling the divide between indigenous and scientific knowledge. Dev Change 26:413–439

Agrawal A (2009) Why 'indigenous' knowledge? J R Soc N Z 39(4):157–158

Agrawal A, Gibson CC (1999) Enchantment and disenchantment: the role of community in natural resource conservation. World Dev 27(4):629–649

Alcamo J, Florke M, Marker M (2007) Future long-term changes in global water resources driven by socio-economic and climatic changes. Hydrol Sci J 52(2):247–275

Alcorn JB (1989) Process as resource. Adv Econ Bot 7:31–63

Alcorn JB (1993) Indigenous peoples and conservation. Conserv Biol 7(2):424–426

Alexander LV, Zhang X, Peterson TC, Caesar J, Gleason B, KleinTank AMG, Haylock M, Collins D, Trewin B, Rahimzadeh F, Tagipour A, Rupa Kumar K, Revadekar J, Griffiths G, Vincent L, Sthepenson DB, Burn J, Aguilar E, Brunet M, Taylor M, New M, Zhai P, Rusticucci M, Vazquez-Aguirre L (2006) Global observed changes in daily climate extremes of temperature and precipitation. J Geophys Res 111:D05109. doi:10.1029/2005JD006290

Anderegg WRL, Prall JW, Harold J, Schneider SH (2010) Expert credibility in climate change. Proc Natl Acad Sci U S A 107(27):12107–12109

Anshelm J, Hansson A (2011) Climate change and the convergence between ENGOs and business: on the loss of utopian energies. Environ Values 20:75–94

Aubertin C (2002) De Rio à Johannesburg, les avatars de la biodiversité. In: Martin J-Y (dir) Développement durable? Doctrines, pratiques et évaluations. IRD, Paris

Aubertin C, Boisvert V, Franck-Dominique V (1998) La construction sociale de la question de la biodiversité. Nat Sci Soc 6(1):7–19

Bakker K, Bridge G (2006) Material worlds? Resource geographies and the 'matter of nature'. Prog Hum Geogr 30(1):5–27

Berkes F (1993) Traditional ecological knowledge in perspective. In: Inglis JT (ed) Traditional ecological knowledge: concepts and cases. International Development Research Centre, Ottawa

Berkes F (2008) Sacred ecology, 2nd edn. Routledge, New York

Berkes F (2009) Indigenous ways of knowing and the study of environmental change. J R Soc N Z 39(4):151–156

Berkes F, Davidson-Hunt IJ (2006) Biodiversity, traditional management systems, and cultural landscapes: examples from the Boreal forest of Canada. Int Soc Sci J 58(187):35–47

Berkes F, Jolly D (2001) Adapting to climate change: social-ecological resilience in a Canadian Western Arctic Community. Conserv Ecol 5(2):18. www.consecol.org/vol5/iss2/art18

Berkes F, Turner NJ (2006) Knowledge, learning and the evolution of conservation practice for social-ecological system resilience. Hum Ecol 34(4):479–494

Berkes F, Colding J, Folke C (2000) Rediscovery of traditional ecological knowledge as adaptive management. Ecol Appl 10(5):1251–1262

Bernauer T, Böhmelt T, Koubi V (2012) Environmental changes and violent conflict. Environ Res Lett 7(1):015601. doi:10.1088/1748-9326/7/1/015601

Bhagwat SA, Rutte C (2006) Sacred groves: potential for biodiversity management. Front Ecol Environ 4(10):519–524

Bonny E, Berkes F (2008) Communicating traditional environmental knowledge: addressing the diversity of knowledge, audiences and media type. Polar Rec 44(230):243–253

Botkin DB, Saxe H, Araújo MB et al (2007) Forecasting the effects of global warming on biodiversity. BioScience 57(3):227–236

Braun B, Castree N (eds) (1998) Remaking reality: nature at the millennium. Routledge, New York

Brown RD (2010) Analysis of snow cover variability and change in Québec 1948–2005. Hydrol Process 24:1929–1954

Budds J (2009) Constested H_2O: science, policy and politics in water resources management in Chile. Geoforum 40:418–430

Butchart SHM, Matt W, Ben C, Arco VS, Scharlemann JPW, Almond REA, Ballie JEM, Bomhard B, Brown C, Bruno J, Carpenter KE, Carr GM, Chanson J, Chenery AM, Csirke J et al (2010) Global biodiversity: indicators of recent declines. Science 328(5982):1164–1168

Caillon S, Degeorges P (2007) Biodiversity: negotiating the border between nature and culture. Biodivers Conserv 16:2919–2931

Calamia MA (1999) Une méthodologie visant à incorporer les connaissances écologiques traditionnelles aux systèmes d'information géographique pour gérer les ressources marines dans le Pacific. Resour Mar Tradit 10:2–12

Carolan MS (2007) Saving seeds, saving culture: a case study of a heritage seed bank. Soc Nat Resour 20(8):739–750

Chapin M (2000) The seduction of models: Chinampa agriculture in Mexico. Grassroots Dev 12 (1):8–17

Christensen JH, Hewitson B, Busuioc A, Chen A, Gao S, Held I, Jones R, Kolli RK, Kwon W-T, Laprise R, Magaña Rueda V, Mearns L, Menéndez CG, Räisänen J, Rinke A, Sarr A, Whetton P (2007) Regional climate projections. In: Solomon S, Qin D, Manning M, Hen Z, Marquis M, Avery KB, Tognor M, Miller HL (eds) Climate change 2007: the physical science basis. Contribution of working group 1 to the fourth assessment report of the intergovernmental panel on climate change. Cambridge University Press, Cambridge

Cline WR (2007) Global warming and agriculture: impact estimates by country. Center for Global Development and the Peterson Institute for International Economic, Washington, DC

Colchester M (2004) Conservation policy and indigenous peoples. Environ Sci Policy 7:145–153

Crawford S, Wehkamp CA, Smith N (2010) Translation of indigenous/western science perspectives on adaptive management for environmental assessments, Research and development monograph series. Canadian Environmental Assessment Agency, Ottawa

Cronin SJ, Gaylord DR, Douglas C, Alloway BV, Wallez S, Esau JW (2004) Participatory methods of incorporating scientific with traditional knowledge for volcanic hazard management on Ambae Island, Vanuatu. Bull Volcanol 66:652–668

Cronon W (1983) Changes in the land: Indians, colonists, and the ecology of New England. Hill & Wang, New York

Cronon W (ed) (1996) Uncommon ground: rethinking the human place in nature. W.W. Norton & Company, New York

Daily GC (1997) Nature's services: societal dependence on natural ecosystems. Island Press, Washington, DC

Davidson DJ, Williamson T, Parkings JR (2003) Understanding climate change risk and vulnerability in northern forest-based communities. Can J Forest Res 33:2252–2261

Davidson-Hunt IJ (2003) Journeys, plants and dreams: adaptive learning and social-ecological resilience. PhD Thesis, University of Manitoba, Canada

Davidson-Hunt Iain J, Berkes Fikret (2003) Learning as you journey: Anishinaabe perception of social-ecological environments and adaptive learning. Conserv Ecol 8(1):art5 www.consecol. org/vol8/iss1/art5

Davis AJ, Jenkinson LS, Lawton JH, Shorrocks B, Wood S (1998a) Making mistakes when predicting shifts in species range in response to global warming. Nature 39:783–786

Davis AJ, Lawton JH, Shorrocks B, Jenkinson LS (1998b) Individualistic species responses invalidate simple physiological models of community dynamics under global environment change. J Anim Ecol 67:600–612

Descola P (2005) Par-delà nature et culture. Gallimard, Paris

Desjarlais C, Allard M, Bélanger D, Blondot A, Bourque A, Chaumont D, Gosselin P, Houle D, Larrivée C, Lease N, Pham AT, Roy R, Savard J-P, Turcotte R, Villeneuve C (2010) OURANOS. Savoir s'adapter aux changements climatiques. OURANOS, Montréal

Diamond J (1993) New Guineans and their natural world. In: Kellert SR, Wilson EO (eds) The biophilia hypothesis. Island Press, Washington, DC

Dickison M (2009) The asymmetry between science and traditional knowledge. J R Soc N Z 39 (4):171–172

Dominique B, de Blois S, Angers J-F et al (2010) The CC-bio project: studying the effects of climate change on Quebec biodiversity. Diversity 2:1181–1204

Downing A, Cuerrier A (2011) A synthesis of the impacts of climate change on the First Nations and Inuit of Canada. Indian J Tradit Knowl 10(1):57–70

Duerden F (2004) Translating climate change impacts at the community level. Arctic 57 (2):204–212

Fast H, Berkes F (1998) Climate change, northern subsistence and land based economies. In: Mayer N, Avis W (eds) The Canada country study: climate impacts and adaptation, volume VIII: national cross-cutting issues volume. Environment Canada, Canada

Flannigan MD, Bergeron Y, Engelmark O, Wotton BM (1998) Future wildfire in circumboreal forests in relation to global warming. J Veg Sci 9:469–476

Fleming JR (1998) Historical perspectives on climate change. Oxford University Press, New York

Folke C, Lowell P Jr, Berkes F, Colding J, Svedin U (2007) The problem of fit between ecosystems and institutions: ten years later. Ecol Soc 12(1):30. www.ecologyandsociety.org/vol12/iss1/ art30

Ford JD (2009) Vulnerability of Inuit food systems to food insecurity as a consequence of climate change: a case study from Igloolik, Nunavut. Reg Environ Chang 9:83–100

Ford JD, Pearce T, Smit B, Wandel J, Allurut M, Shappa K, Ittusujurat H, Qrunnut K (2007) Reducing vulnerability to climate change in the Arctic: the case of Nunavut, Canada. Arctic 60 (2):150–166

Forsyth T (2003) Critical political ecology: the politics of environmental science. Routledge, London

Fourier J-BJ (1888) Théorie analytique de la chaleur. Gauthier-Villars et fils, Paris

Fraser DJ, Coon T, Prince MR, René D, Bernatchez L (2006) Integrating traditional and evolutionary knowledge in biodiversity conservation: a population level case study. Ecol Soc 11(2):4. www.ecologyandsociety.org/vol11/iss2/art4/

Freeman MMR (1992) The nature and utility of traditional ecological knowledge. North Perspect 20(1):9–12

Froyd CA, Willis KJ (2008) Emerging issues in biodiversity & conservation management: the need for a paleoecological perspective. Quat Sci Rev 27:1723–1732

Furgal C, Prowse TD (2008) Northern Canada. In: Lemmen DS, Warren FJ, Lacroix J, Bush E (eds) From impacts to adaptation: Canada in a changing climate 2007. Government of Canada, Ottawa

Furgal C, Seguin J (2006) Climate change, health and vulnerability in Canadian northern aboriginal communities. Environ Health Perspect 114(12):1964–1970

Gagnon AS, Gough WA (2005) Trends in the dates of ice freeze-up and breakup over Hudson Bay, Canada. Arctic 58(4):370–382

Gari J-A (1999) Biodiversity conservation and use: local and global considerations. Science, Technology and Development Discussion Paper No.7, centre for International Development and Belfer Center for Science and International Affairs. Harvard University, Cambridge

Gearheard S, Matumeak W, Angutikjuaq i, Malanik JP, Huntinton H, Leavitt J, Matumeak Kagak D, Tigullaraq G, Barry RG (2006) "It's not that simple" a collaborative comparison of sea ice environments, their uses, observed changes, and adaptations in Barrow, Alaska, USA, and Clyde River, Nunavut, Canada. AMBIO J Hum Environ 35(4):203–211

George P, Preston RJ (1987) 'Going in Between': the impact of European technology on the work patterns of the West Main Cree of Northern Ontario. J Econ Hist 47:447–460

George P, Berkes F, Preston RJ (1995) Aboriginal harvesting in the Moose River Basin: a historical and contemporary analysis. Can Rev Sociol Anthropol 32:69–90

Girardin MP, Mudelsee M (2008) Past and future changes in Canadian boreal wildfire activity. Ecol Appl 18(2):391–406

GCOS – Global Climate Observing System (2003) The second report on the adequacy of the global observing systems for climate in support of the UNFCCC, GCOS-82, WMO/TD No. 1143. World Meteorological Organization

Gough WA (1998) Projections of sea-level change in Hudson and James Bays, Canada, due to global warming. Arct Alp Res 30(1):84–88

Greene S (2004) Indigenous people incorporated? Culture as politics, culture as property in pharmaceutical bioprospecting. Curr Anthropol 45(2):211–237

Gumuchian H (1990) À la périphérie de la périphérie: l'espace rural et le concept de fragilité en Abitibi, Collection Notes et documents, 90-01. Université de Montréal, Département de géographie, Montréal

Gupta A (2009) Geopolitical implications of Arctic meltdown. Strateg Anal 33(2):174–177

Hansen JE (1988) The greenhouse effect: impacts on current global temperature and regional heat waves. Statement presented to: U.S. Senate Committee on Energy and Natural Resources

Hansen James E (2005) A slippery slope: how much global warming constitutes "dangerous anthropogenic interference"? Clim Change 68:269–279

Hansen J, Sato M, Ruedy R, Lacis A, Oinas V (2000) Global warming in the twenty-first century: an alternative scenario. Proc Natl Acad Sci 97(18):9875–9880

Hansen J, Nazarenko L, Ruedy R, Sato M, Willis J, Del Genio A, Koch D, Lacis A, Lo K, Menon S, Novakov T, Perlwitz J, Russell G, Schmidt GA, Tausnev N (2005) Earth's energy imbalance: confirmation and implications. Science 308:1431–1435

Haraway D (1989) Primate visions: gender, race and nature in the world of modern science. Routledge, New York

Hely C, Girardin MP, Ali AA, Carcaillet C, Brewer S, Bergeron Y (2010) Eastern boreal North American wildfire risk of the past 7000 years: a model-data comparison. Geophys Res Lett 37: L14709. doi:10.1029/2010GL043706

Herrmann TM, Martin GJ, Laxmi P, Borrini-Feyerabend G, Hay-Edie T, Oldham P, Dutfield G (2010) Biocultural diversity and development under local and global change. In: Ibisch PL, Vega E, Herrmann TM (eds) Interdependence of biodiversity and development under global change, Technical series no. 54. Secretariat of the Convention on Biological Diversity, Montréal

Herrmann TM, Royer M-JS, Cuciurean R (2012) Understanding subarctic wildlife in Eastern James Bay under changing climatic and socio-environmental conditions: bringing together Cree hunters' ecological knowledge and scientific observations. Polar Geogr 35(3–4):245–270

Higgins C (2000) Indigenous knowledge and occidental science: how both forms of knowledge can contribute to an understanding of sustainability. In: Hollstedt C, Sutherland K, Innes T (eds) Proceedings from science to management and back: a science forum for Souther Interior Ecosystems of British Columbia. Southern Interior Forest Extension and Research Partnership, Kamloops

Houde N (2007) The six faces of traditional ecological knowledge: challenges and opportunities for Canadian co-management arrangements. Ecol Soc 12(2):34. www.ecologyandsociety.org/vol12/iss2/art34

Houghton JT, Jenkins GJ, Ephraums JJ (eds) (1990) Climate change: the IPCC scientific assessment, report prepared for IPCC by Working Group I. Cambridge University Press, New York

Huntington HP (1998) Observations on the utility of the semi-directive interview for documenting traditional ecological knowledge. Arctic 51(3):237–242

Huntington HP (2000) Using traditional ecological knowledge in science: methods and applications. Ecol Appl 10(5):1270–1274

IPCC – Intergovernmental Panel on Climate Change (2007a) Contribution of working Groups I, II and III to the fourth assessment report of the intergovernmental panel on climate change [Core Writing Team, Pachauri RK and Reisinger A (eds)]. IPCC, Geneva

IPCC – Intergovernmental Panel on Climate Change (2007b) Climate change 2007: the physical science basis. In: Solomon S, Qin D, Manning M, Chen Z, Marquis M, Averyt KB, Tignor M, Miller HL (eds) Contribution of working group I to the fourth assessment report of the Intergovernmental Panel on Climate Change. Cambridge University Press, Cambridge

IPCC – Intergovernmental Panel on Climate Change (2013a) Principle governing IPCC work. IPCC, Geneva

IPCC – Intergovernmental Panel on Climate Change (2013b) Climate change 2013: the physical science basis. In: Stocker TF, Qin D, Plattner G-K, Tignor M, Allen SK, Boschung J, Nauels A, Xia Y, Bex V, Midgley PM (eds) Contribution of working group I to the fifth assessment report of the intergovernmental panel on climate change. Cambridge University Press, Cambridge

IPCC – Intergovernmental Panel on Climate Change (2014) Summary for policymakers. In: Field CB, Barros VR, Dokken DJ, Mach KJ, Mastrandrea MD, Bilir TE, Chatterjee M, Ebi KL, Estrada YO, Genova RC, Girma B, Kissel ES, Levy AN, MacCracken S, Mastrandrea PR, White LL (eds) Climate change 2014: impacts, adaptation and vulnerability. Part A: global and sectoral aspects. Contribution of working group II to the fifth assessment report of the Intergovernmental Panel on Climate Change. Cambridge University Press, Cambridge

Jacqmain H, Bélanger L, Hilton S, Bouthillier L (2007) Bridging native and scientific observations of snowshoe hare habitat restoration after clearcutting to set wildlife habitat management guidelines on Waswanipi Cree land. Can J Forest Res 37:530–539

Jacqmain H, Délanger L, Courtois R, Dussault C, Beckley TM, Pelletier M, Gull SW (2012) Aboriginal forestry: development of a socioecologically relevant moose habitat management process using local Cree and scientific knowledge in Eeyou Istchee. Can J Forest Res 42:631–641

Johnson LM (1999) Aboriginal burning for vegetation management in Northwest British Columbia. In: Boyd R (ed) Indians, fire and the land in the Pacific Northwest. Oregon State University Press, Oregon

Jones PD, Moberg A (2003) Hemispheric and large-scale surface air temperature variations: an extensive revision and an update to 2001. J Climate 16:206–223

Kobashi T, Severinghaus JP, Brook EJ, Barnola J-M, Grachev AM (2000) Precise timing and characterization of abrupt climate change 8200 years ago from air trapped in polar ice. Quat Sci Rev 26:1212–1222

Kumar GS, McKey d, Aumeeruddy-Thomas Y (2004) Heterogeneity in ethnoecological knowledge and management of medicinal plants in the Himalayas of Nepal: implications for conservation. Ecol Soc 9(3):6. www.ecologyandsociety.org/vol9/iss3/art6/

Lafortune V, Filion L, Hétu B (2006) Impacts of Holocene climatic variations on alluvial fan activity below snow patches in Subarctic Québec. Geomorphology 76:375–391

Laidler GJ (2006) Inuit and scientific perspectives on the relationship between sea ice and climate change: the ideal complement? Clim Change 78:407–444

Latour B (1987) Science in action: how to follow scientists and engineers through society. Harvard University Press, Cambridge

Latour B (1991) Nous n'avons jamais été modernes: essai d'anthropologie symétrique. Éditions La Découverte, Paris

Lavoie C, Allard M, Duhamel D (2012) Deglaciation landforms and C-14 chronology of the Lac Guillaume-Delisle area, Eastern Hudson Bay: a report on field evidence. Geomorphology 159–160:142–1555

Lehner B, Döll P, Alcamo J, Henrichs T, Kaspar F (2006) Estimating the impact of global change on flood and drought risks in Europe: a continental, integrated analysis. Clim Change 75:273–299

Lemmen DS, Warren FJ, Lacroix J, Bush E (eds) (2008) From impacts to adaptation: Canada in a changing climate 2007. Government of Canada, Ottawa

Linden E (1991) Lost tribes, lost knowledge. Time 38(12):44–58

Liverman DM (2008) Conventions of climate change: constructions of danger and the dispossession of the atmosphere. J Hist Geogr. doi:10.1016/j.jhg.2008.08.008

Lyver PO'B, Jones C, Moller H (2009) Looking past the wallpaper: considerate evaluation of traditional environmental knowledge by science. J R Soc N Z 39(4):219–223

MacCraken MC, Luther FM (eds) (1985) Detecting the climatic effects of increasing carbon dioxide. United States Department of Energy, Washington, DC

Maffi L (ed) (2001) On biocultural diversity: linking language, knowledge, and the environment. Smithsonian Institute Press, Washington, DC

Maffi L (2005) Linguistic, cultural, and biological diversity. Annu Rev Anthropol 34:599–617

Maffi L, Woodley E (2010) Biocultural diversity conservation: a global sourcebook. Earthscan, Washington, DC

Martin E (1991) The egg and the sperm: how science has constructed a romance based on stereotypical male-female roles. Signs 16(3):485–501

Mayer N, Avis W (eds) (1998) The Canada country study: climate impacts and adaptation, volume VIII: national cross-cutting issues volume. Environment Canada, Canada

Mekis É, Hogg WD (1999) Rehabilitation and analysis of Canadian daily precipitation time series. Atmos Ocean 37(1):53–85

Mekis É, Vincent LA (2011) An overview of the second generation adjusted daily precipitation dataset for trend analysis in Canada. Atmos Ocean 49(2):163–177

Melillo JM, McGuire DA, Kicklighter DW, Moore B III, Vorosmarty CJ, Schloss AL (1993) Global climate change and terrestrial net primary production. Nature 363:234–240

Menzies CR (2006) Traditional ecological knowledge and natural resource management. University of Nebraska Press, Nebraska

Millennium Ecosystem Assessment (MEA) (2005) Ecosystems and human well-being: biodiversity synthesis. World Resources Institute, Washington, DC

Mitchell WL, Brown PF (2002) Les populations de montagne – adaptation et persistance d'une culture, à l'aube d'un nouveau siècle. Unasylva 208(53):47–53

Mitrovica JX, Forte AM, Simons M (2000) A reappraisal of postglacial decay times for Richmond Gulf and James Bay, Canada. Geophys J Int 142:783–800

Moller H, Berkes F, O'Brian Lyver P, Kislalioglu M (2004) Combining science and traditional ecological knowledge: monitoring populations for co-management. Ecol Soc 9(3):2. www.ecologyandsociety.org/vol9/iss3/art2

Moser SC (2010) Now more than ever: the need for more societally relevant research on vulnerability and adaptation to climate change. Appl Geogr 30:464–474

Mühlhäusler P (1995) The interdependence of linguistic and biological diversity. In: Myers D (ed) The politics of multiculturalism in the Asia/Pacific. North Territory University Press, Darwin

Murray RW (2012) Arctic politics in the emerging multipolar system: challenges and consequences. Polar J 2(1):7–20

Nadasdy P (2005) Transcending the debate over the ecologically noble Indian: indigenous peoples and environmentalism. Ethnohistory 52(2):291–331

Nakashima DJ (1993) Astute observers on the sea ice edge: Inuit knowledge as a basis for Arctic co-management. In: Inglis JT (ed) Traditional ecological knowledge: concepts and cases. IDRC, Ottawa

Nakashima DJ, Roué M (2002) Indigenous knowledge, peoples and sustainable practice. In: Munn T (ed) Encyclopaedia of global environmental change. Wiley, Chinchester

NASCSCC – National Academy of Sciences Committee on the Science of Climate Change (2001) Climate change science: an analysis of some key questions. National Academy Press, Washington, DC

Nevins HM, Adams J, Moller H, Newman J, Hester M, Hyrenbach DM (2009) International and cross-cultural management in conservation of migratory species. J R Soc N Z 39(4):183–185

Nichols T, Berkes F, Jolly D, Snow NB (2004) The community of Sachs Harbour, climate change and sea ice: local observations from the Canadian Western Arctic. Arctic 57(1):68–79

O'Brien K, Sygna L, Haugen JE (2004) Vulnerable or resilient? A multi-scale assessment of climate impacts and vulnerability in Norway. Clim Change 64:193–225

O'Brien K, Eriksen S, Sygna L, Naess LO (2006) Questioning complacency: climate change impacts, vulnerability, and adaptation in Norway. Ambio 35(2):50–56

O'Flaherty Michael R, Davidson-Hunt Lain J, Manseau M (2008) Indigenous knowledge and values in planning for sustainable forestry: Pikangikum First Nation and the Whitefeather forest initiative. Ecol Soc 13(1):6. www.ecologyandsociety.org/vol13/iss1/art6/

O'Neill BC, Oppenheimer M (2004) Climate change impacts are sensitive to the concentration stabilization path. Proc Natl Acad Sci 101(47):16411–16416

Ohl C, Johst K, Meyerhoff J, Beckenkamp VG, Drechsler M (2010) Long-term socio-ecological research (LTSER) for biodiversity protection – a complex systems approach for the study of dynamic human-nature interactions. Ecol Complex 7:170–178

Oreskes N (2004) The scientific consensus on climate change. Science 306:1686

Patenaude G, Reveret JP, Potvin C (2002) Perspectives locales sur les priorités de conservation faunique. VertigO 3(1): http://vertigo.revues.org/4130

Payette S, Fortin M-J, Gamache I (2001) The subarctic forest-tundra: the structure of a biome in a changing climate. BioScience 51(9):709–718

Payette S, Boudreau S, Morneau C, Pitre N (2004) Long-term interactions between migratory caribou, wildfires and nunavik hunters inferred from tree rings. Ambio 33(8):482–486

Pearce TD, Ford JD, Laidler GJ, Smit B, Duerden F, Allarut M, Andrachuk M, Dialla A, Elee P, Goose A, Ikummaq T, Joamie E, Kataoyak F, Loring E, Meakin S, Nickels S, Shappa K, Shirley J, Wandel J (2009) Community collaboration and climate change research in the Canadian Arctic. Polar Res 28:10–27

Pearson RG, Dawson TP (2003) Predicting the impacts of climate change on the distribution of species: are bioclimate envelope models useful? Glob Ecol Biogeogr 12:361–371

Peloquin C, Berkes F (2009) Local knowledge, subsistence harvest and social-ecological complexity in James Bay. Hum Ecol 37:533–545

Peterson TC, Eaterling DR, Karl TR, Goisman P, Nicholls N, Plummer N, Torok S, Auer I, Oehm R, Gullett D et al (1998) Homogeneity adjustments of *in situ* atmospheric climate data: a review. Int J Climatol 18:1493–1517

Pinton F (2003) Traditional knowledge and areas of biodiversity in Brazilian Amazonia. Int Soc Sci J 55(178):607–618

Plummer R, Armitage D (2007) Crossing boundaries, crossing scales: the evolution of environment and resource co-management. Geogr Compass 1(4):834–849

Posey DA (ed) (1999) Cultural and spiritual values of biodiversity. UNEP/Intermediate Technology Publications, London

Posey DA (2002) Commodification of the sacred through intellectual property rights. J Ethnopharmacol 83(1–2):3–12

Preston RJ (2002) Cree narrative: expressing the personal meanings of events, 2nd edn. McGill-Queen's University Press, Montreal

Räisänen J (2008) Warmer climate: less or more snow? Climate Dynam 30:307–319

Raleigh C, Urdal H (2007) Climate change, environmental degradation and armed conflict. Polit Geogr 26(6):674–694

Redford KH, Stearman AM (1993) Forest-dwelling native Amazonians and the conservation of biodiversity. Conserv Biol 7:248–255

Reveret J-P, Charron I, St-Arnaud R-M (2008) Réflexions sur les methods d'estimation de la valeur économique des pertes d'habitats fauniques. Groupe Agéco pour le Ministère des Ressources Naturelles et de la Faune, Direction du développement socio-économique, des partenariats et de l'éducation, Québec

Richardson B (1976) Strangers devour the land. Alfred A Knopf, New York

Riedlinger D, Berkes F (2001) Contributions of traditional knowledge to understanding climate change in the Canadian Arctic. Polar Rec 37(203):315–328

Rosenweigh C, Parry ML (1994) Potential impacts of climate change on world food supply. Nature 367:133–138

Rothman DS, Demeritt D, Chiotti Q, Burton I (1998) Costing climate change: the economics of adaptations and residual impacts for Canada. In: Mayer N, Avis W (eds) The Canada country study: climate impacts and adaptation, volume VIII: national cross-cutting issues volume. Environment Canada, Canada

Roué M, Nakashima D (2002) Des savoirs 'traditionnels' pour évaluer les impacts environnementaux du développement moderne et occidental. Rev Int Sci Soc 3(173):377–387

Rousse J (1987) Knowledge and power: toward a political philosophy of science. Cornell University Press, New York

Roussel B (2005) Savoirs locaux et conservation de la biodiversité: renforcer la représentation des communautés. Mouvements 41:82–88

Samson J, Berteaux D, McGill BJ, Humphries MM (2011) Geographic disparities and moral hazards in the predicted impacts of climate change on human population. Glob Ecol Biogeogr. doi:10.1111/j.1466-8238.2010.00632.x

Saucan DR (1999) Le développement durable et les zones rurales fragiles au Québec. Mémoire présenté à la faculté des études supérieures en vue de l'obstention du grad de Maître ès sciences, Département de géographie, Faculté des arts et des sciences, Université de Montréal, Montréal

SCDB – Secretary of the Convention on Biological Diversity (2006) Global biodiversity outlook 2. Montreal

Scott C (1989) Knowledge construction among Cree hunters: metaphors and literal understanding. J Soc Am 75:192–208

Secretary-General of the United nations (UNSG) (1993) Convention on biological diversity. Concluded at Rio de Janeiro on 5 June 1992. UNEP, Geneva

Shabecoff P (1988) Global warming has begun, expert tells senate. New York Time, New York

Singh-Negi C (2010) Traditional culture and biodiversity conservation: examples from Uttarankhand, Central Himalaya. Mt Res Dev 30(3):259–265

Stephenson J, Moller H (2009) Cross-cultural environmental research and management: challenges and progress. J R Soc N Z 39(4):139–149

Stevenson M (1996) In search of inuit ecological knowledge: a protocol for its collection, interpretation and use: a discussion Paper. Report to the Department of Renewable Resources, GNWT, Qiqiqtaaluk Willife Board and Parks Canada, Iqaluit, Nunavut

Sutherland WJ (2003) Parallel extinction risk and global distribution of languages and species. Nature 423:276–279

Toledo VM (1992) What is ethnoecology? Origins, scope and implications of a rising discipline. Etnoecológica 1(1):5–21

Toledo VM (2001) Indigenous peoples and biodiversity. In: Levin S et al (eds) Encyclopaedia of biodiversity. Academic, San Diego

Tremblay M, Furgal C, Lafortune V, Larrivée C, Savard J-P, Barrett M, Annanack T, Enish N, Tookalook P, Etidloie B (2006) Communities and ice: bringing together traditional and scientific knowledge. In: Riewe R, Oakes J (eds) Climate change: linking traditional and scientific knowledge. Aboriginal Issues Press, University of Manitoba, Manitoba

Trenberth KE, Jones PD, Ambenje R, Bojariu R, Easterling D, Klein Tank A, Parker D, Rahimzadeh F, Renwick JA, Rusticci M, Soden B, Zhai P (2007) Observations: surface and atmospheric climate change. In: Solomon S, Qin D, Manning M, Hen Z, Marquis M, Avery KB, Tognor M, Miller HL (eds) Climate change 2007: the physical science basis. Contribution of working group 1 to the fourth assessment report of the Intergovernmental Panel on Climate Change. Cambridge University Press, Cambridge

Tsuji LJS, Nieboer E (1999) A question of sustainability in Cree harvesting practices: the seasons, technological and cultural changes in the Western James Bay region of Northern Ontario, Canada. Can J Nativ Stud 19(1):169–192

Turner NJ (1999) 'Time to burn'. Traditional use of fire to enhance resource production by aboriginal peoples in British Columbia. In: Boyd R (ed) Indians, fire and the land in the Pacific Northwest. Oregon State University Press, Oregon

Turner NJ, Clifton H (2009) 'It's so different today': climate change and indigenous lifeways in British Columbia, Canada. Glob Environ Chang 19:180–190

Turner NJ, Ignace MB, Ignace R (2000) Traditional ecological knowledge and wisdom of aboriginal peoples in British Columbia. Ecol Appl 10(5):1275–1287

Tyrrell M (2006) 'More bears, less bears': Inuit and scientific perceptions of polar bear populations on the west coast of Hudson Bay. Inuit Stud 30(2):191–208

UNESCO (2003) Diversité culturelle et biodiversité pour un développement durable, Table ronde de haut niveau organisée conjointement par l'UNESCO et le PNUE le 3 septembre 2002 à Johannesburg (Afrique du Sud) à l'occasion du Sommet mondial pour le développement durable. Organisation des Nations Unies pour l'éducation, la science et la culture, Paris

UNESCO (2009a) Continuité et changement: le dynamisme des savoirs 'traditionnels'. Links, Paris

UNESCO (2009b) Savoirs locaux et développement durable. Links, Paris

United Nations (1992) United Nations framework convention on climate change. United Nations, New York

Usher PJ (2000) Traditional ecological knowledge in environmental assessment and management. Arctic 53(2):183–193

Vallée S, Payette S (2007) Collapse of permafrost mounds along a subarctic river over the last 100 years (Northern Québec). Geomorphology 90:162–170

Van Eijck M, Roth W-M (2007) Keeping the local local: recalibrating the status of science and traditional ecological knowledge (TEK) in education. Sci Educ 91:926–947

Vincent LA, Zhang X, Bonsal BR, Hogg WD (2002) Homogenization of daily temperatures over Canada. J Climate 15:1322–1334

Wainwright J (2010) Climate change, capitalism, and the challenge of transdisciplinarity. Ann Assoc Am Geogr 100(4):983–991

Warren DM, Von Liebenstein G, Slikkerveer L (1993) Networking for indigenous knowledge. Indigen Knowl Dev Monit 1(1):2–4

Watson A, Alessa L, Glaspell B (2008) The relationship between traditional ecological knowledge, evolving cultures and wilderness protection in the circumpolar North. Conserv Ecol 8 (1):2. www.consecol.org/vol8/iss1/art2

Waylen KA, Fischer A, McGowan PJK, Thirgood SJ, Milner-Gulland EJ (2010) Effect of local cultural context on the success of community-based conservation interventions. Conserv Biol 24(4):1119–1129

Weatherhead E, Gearheard S, Barry RG (2010) Changes in weather persistence: insight from Inuit knowledge. Glob Environ Chang 20:523–528

Weaver AJ (2004) The science of climate change. In: Coward H, Weaver AJ (eds) Hard choices: climate change in Canada. Wilfrid Laurier University Press, Waterloo

Wenzel GW (1999) Traditional ecological knowledge and Inuit: reflections on TEK research and ethics. Arctic 52(2):113–124

Western D, Strum SC, Wright MR (eds) (1994) Natural connections: perspectives in community-based conservation. Island Press, Washington, DC

White R (1995) The organic machine: the remaking of the Columbia River. Hill and Wang, New York

White R (1996) 'Are you an environmentalist or do you work for a living?': work and nature. In: Cronon W (ed) Uncommon ground: rethinking the human place in nature. W.W. Norton & Company, New York

Woodward IF, Beerling DJ (1997) The dynamics of vegetation change: health warning for equilibrium 'dodo' models. Glob Ecol Biogeogr Lett 6:413–418

World Commission on Environment and Development (WCED), Brundtland Gro Harlem (chairman) (1987) Our common future. Oxford University Press, Oxford

Yagouti A, Boulet G, Vincent L, Vescovi L, Mekis É (2008) Observed changes in daily temperatures and precipitation indices for southern Québec, 1960–2005. Atmos Ocean 46(2):243–256

Zhang X, Vincent LA, Hogg WD, Niitsoo A (2000) Temperature and precipitation trends in Canada during the 20th century. Atmos Ocean 38(3):395–429

Zhang X, Hegerl G, Zwiers FW, Kenyon J (2005) Avoiding inhomogeneity in percentile-based indices of temperature extremes. J Climate 18:1641–1651

Chapter 3
Eastern James Bay and the Cree

3.1 The Eastern James Bay

3.1.1 Borders and Administration

Up until 1870 when it was transferred to the Dominion of Canada by the United Kingdom, the region encompassing the territory surrounding the James and Hudson Bays and part of northern Quebec belonged to the Hudson Bay Company and was known as Rupert's Land. In 1898 the Parliament of Canada fixed the northern border of the Quebec province at the middle of the Eastmain River. Everything north of this border belonged to the Ungava district of the North-West-Territories. In 1912 the Ungava district is transferred to Quebec. Then in 1927 its eastern border, between Quebec and Newfoundland, is set by the Judicial Committee of the Privy Council in London. In 1966 the New Quebec region is created and is comprised of all the land north of the 50th parallel. Finally in 1987 New Quebec's territory is expanded to the south until the 49th parallel and is renamed Nord-du-Québec administrative region (AR).

The Nord-du-Québec AR is composed of two regional administration territories: the James-Bay (inhabited by the Cree) and Kativik (which includes the Inuit land of Nunavik and the Naskapi village of Kawawachikamach). It is Quebec's largest administrative region, covering 55 % of the province (860,553 km^2) and also it least populated area with only 0.5 % of the province's population living there (41,479 people in 2009) (MDDEPQ 2011a; MDECRQ 2011). The James Bay Territory, named Eeyou Istchee by the Cree, represents the southern part of the region, extending from the 49th to the 55th parallel and from the James Bay to the west until the 68th western meridian at its eastern limit (Ducruc et al. 1976). It covers 297,330 km^2 which represents 20 % of Quebec's land surface area (MBJ 2010; CRBJ 2011).

There are five Quebec municipalities, with a combined population of 15,351 inhabitants in 2011, located on the category III lands (see conventions and

© The Author(s) 2016
M.-J.S. Royer, *Climate, Environment and Cree Observations*,
SpringerBriefs in Climate Studies, DOI 10.1007/978-3-319-25181-3_3

agreements) of the James Bay Territory (MAMROT 2011). They are: Chapais (established in 1955), Chibougamau (established in 1952), Lebel-sur-Quévillon (established in 1965), Matagami (established in 1963) and Gouvernement regional d'Eeyou Istchee Baie-James (established in 1971; previously called Baie-James municipality) (MBJ 2011; MAMOT 2014). The Gouvernement regional d'Eeyou Istchee Baie-James municipality includes the communities of Villebois (founded in 1935–1936), Radisson (founded in 1974) and Valcanton (created in 2001 following the merging of the Val-Paradis and Beaucanton localities, both founded in 1935–1936) (MBJ 2010).

Aside from the Quebec municipalities, there are nine Cree villages located on category I lands of the James Bay Territory (Fig. 3.1). These are: Chisasibi, Eastmain, Mistissini, Nemaska, Oujé-Bougoumou, Waskaganish, Waswanipi, Wemindji, and Whapmagoostui. A tenth community, Washaw Sibi was at the time of this research in the process of being recognised as part of Eeyou Istchee; however it is located in the Abitibi-Témiscamingue AR, its members currently living in the Algonquin reserve of Pikogan (CWGPN 2011; WSE 2011). This community was excluded from our study because of its location and also because it hadn't yet been officially recognised by the Cree administration (i.e. Cree Nation Government, Cree Trapper's Association, Grand Council of the Cree) or by the Canadian and Quebec governments.

Each Cree community (also identified as First nation) has its own local government which manages the village and lands that are attached to it. The Grand Council of the Cree (GCC) and the Cree Nation Government (CNG; previously called the Cree Regional Authority) compose the Cree regional government. The GCC was created in 1974 during the negotiations on the James Bay hydroelectric project and is the Cree government's political body; while the CNG was created in 1978 to be its administrative arm. The Cree are the only First Nation group living on the James-Bay territory with 16,350 members in 2011 (Statistics Canada 2012a, b, c, d, e, f, g, h, i).

3.1.2 Conventions and Agreements

Northern Quebec was for a long time ignored by provincial and federal authorities (Scott 1989; Canobbio 2009). Between 1867 and 1960 the federal government is the one in charge of political and judicial services for the region's indigenous populations (Goudreau 2003). This presence is assured by a handful of nursing stations and schools as well as annual visits (starting in the twentieth century, and each visit lasting a few days) of doctors and police officers from the Royal Canadian Mounted Police. Between 1920 and 1950, these teams would crisscross the region irregularly, this service only becoming regular in 1950. The provincial government therefore considered the ply of these populations as a minor problem that was being looked after by the federal government (Goudreau 2003).

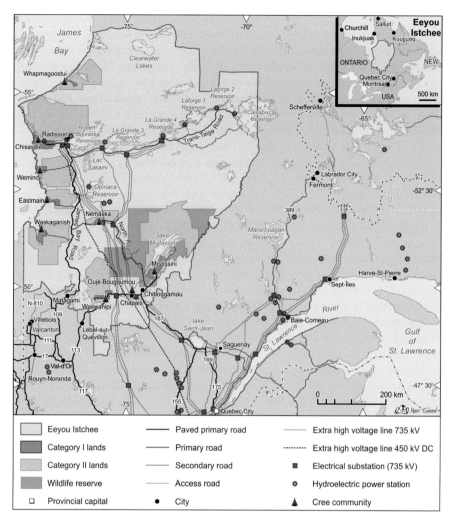

Fig. 3.1 James Bay Territory/Eeyou Istchee (Created and adapted by Marc Girard from Royer 2012)

Things change in 1963 when the provincial government starts seeing the potential offered by the region and creates the "Direction générale du Nouveau-Québec" (DGNQ); which would be responsible for the coordination of activities in northern Quebec (Hamelin 1998; Petit 2010). Following the 1966 election a new party takes power and decides to change the politics undertaken by the previous government in this area. The situation was such that is was described quite dramatically by M. Ciacca in 1975: "Until now, the native peoples have lived, legally speaking, in a kind of limbo. The limits of federal responsibility were never quite clear, nor was it quite clear that Québec had any effective jurisdiction." (SAA 1998, p. XV) No treaty or agreement had ever been signed with the Cree of Eeyou Istchee.

Part of the new vision of the provincial government, under Prime Minister Robert Bourrassa, in the 1970s for the North is the development of a large scale hydroelectric complex in the region. This is in-line with the vision of many other provincial governments of 1960s North America where they "[. . .] thought they could not go wrong in booming, postwar North America by rushing to build mega-dams with mega-loans from their American neighbours" (Froschauer 1999, p. 4). They hoped to take advantage of the United States' high electric demands to develop their provinces' economy. This mind-set was amplified following the Oil Crisis of the 1970s where the price of natural gas and petrol increased exponentially. The whole world was now holding up hydroelectricity as an alternative to these non-renewable resources (White 1995; Swyngedouw 2007). In Quebec this project also follows the Quiet Revolution and Jean Lesage's campaign titled 'Masters in Our Own Home (Maître chez nous); which aimed for control of the land, the territory and the development of the region by the francophone majority (Hogue et al. 1979). "James Bay occupied an important place in this imaging/ imagining of the national territory because it offered a concrete representation of its nature – both symbolic and real – and more specifically of what such a space contained" (Desbiens 2004a, p. 106). The hydro-electric development project for the James Bay is thus clearly inscribed in the political movements of the time.

The project's official announcement takes the Cree and the Inuit by surprise as they hadn't been consulted (Chaplier 2006; Savard 2009). Following this surprise announcement, the Cree mobilised and undertook demonstrations and appeals to the United Nations councils and to the courts of law. One of the appeals was in front of the Judge Malouf who placed an injunction against the work already started on the territory. Although that injunction was overturned a few days later by the Quebec court of appeals, it helped to legitimate the Cree's demands in the eyes of the government (Vincent 1992). These appeals and procedures will culminate with the signing of the James Bay and Northern Quebec Agreement (JBNQA) between the Cree and Inuit representatives, the government of Québec and the government of Canada in 1975 (CCRCCN 1991). In 1978 the Naskapis of Kawachikamach join the JBNQA by adhering to the North Eastern Quebec Agreement (NEQA). The JBNQA applies to the whole of the Nord-du-Québec AR and to part of the lands of the Abitibi administrative region for a total land area of 1,081,000 km^2 (SAA 1998). However different administrative and environmental management rules apply depending on whether you are situated north or south of the 55th parallel, separating the Kativik and Baie-James sectors (MDDEPQ 2011a, b). As our interests lie within the Baie-James sector, south of the 55th parallel, we will concentrate on its modalities.

The JBNQA and the NEQA that modifies it represent the first modern treaty on global land claims in Quebec. Both agreements are considered land claims as defined by section 35(3) of the constitutional law of 1982. As such all institutions put in place are protected by the constitution. The JBNQA includes many gains for the indigenous populations of the Nord-du-Québec AR, among which a financial compensation of about 234 millions of dollars. It also gives them access to important administrative powers (although they must answer to provincial authorities) and puts in place the bases of the administrative framework for the Cree and

Inuit Nations (Le Blanc 2009). The Agreement touches on many aspects of daily life including individual rights, healthcare and education, environmental protection, the administration of justice, social and economic development, and local and regional governments (CCRCCN 1991). In return, the Quebec government obtains the right to develop the region's hydroelectric, mining and forestry resources, as well as the recognition of its sovereignty over the region, with the agreement's articles 2.1 and 2.6 specifying that the Cree and Inuit 'abandon, renounce and cede' all other claims, rights, titles and interests linked to the Nord-du-Québec AR (Mercier and Ritchot 1997; SAA 1998; Savard 2009).

The JBNQA divides the Nord-du-Québec AR into three land type categories (I, II and III). Category I lands are reserved for the exclusive use and benefit of the indigenous populations and includes the indigenous communities' villages. It represents 1.3 % (14,348 km^2) of the convention's territory (MDDEPQ 2011b). For the Cree this represents 5,551.7 km^2 (Transport Québec 2011). The communities and the category I lands are managed by a council put in place through the JBNQA, they do not represent reserves per se as defined by the Indian Act, i.e.:

> Subject to this Act, reserves are held by Her Majesty for the use and benefit of the respective bands for which they were set apart, and subject to this Act and to the terms of any treaty or surrender, the Governor in Council may determine whether any purpose for which lands in a reserve are used or are to be used is for the use and benefit of the band. (Minister of Justice 2015)

When they were created, category I lands represented a new concept; they were lands where the indigenous communities were established, that belonged to the communities and that were created based on traditional activities but remain accessible to the rest of the population (SAA 1998). The structure and powers of the local Cree administrations are detailed in Part I of the Cree-Naskapi (of Quebec) Act. Since 1984 this act replaces the Indian Act for all indigenous people adhering to the JBNQA except concerning Registered Indian status (AADNC 2009). This act addresses various disposition related to local administration and management of the Category I lands (more precisely category IA and IA-N) contained in the JBNQA and the NEQA.

Category II lands are provincial lands on which indigenous populations have exclusive hunting, fishing and trapping rights. They are located in proximity to the villages and represent 14.8 % (159,880 km^2) of the territory covered by the convention, and 70,000 km^2 of lands related to the Cree (MDDEPQ 2011b; The Canadian Encyclopaedia 2012).

Category III lands are provincial public lands and represent the majority of the convention's territory (83.9 % or 907,772 km^2; MDDEPQ 2011b). On these lands, the indigenous population have a right to hunt, fish and trap and 22 species are reserved for their exclusive use. These species include beavers, polar bears, lynxes, foxes, sturgeons, bass and all members of the Mustelidae family (SAA 1998). As public lands, these are accessible to the general public and use of the territory by indigenous and non-indigenous alike must conform to the laws and regulations governing Quebec's public lands (SAA 1998). It is also on these lands that are located the five Quebec municipalities.

The signing of the agreement will however not be the end of the contestations and debates. Over time questions have risen about the way the various promises were upheld and the follow through in putting in place some of the more important clauses from the original accords by the provincial and federal governments (SAA 1998). The complexity of the JBNQA makes it prone to conflict. It required the passing of a federal legislation and over 20 provincial laws to put it into full application (La Rusic 1979). In 2002, a new agreement between the Cree and the Quebec government, the *Agreement Concerning a New Relationship between Le Gouvernement du Québec and the Crees of Québec*, also known as the Peace of the Braves/Paix des Braves, looked at solving many of the conflictual elements (Desbiens 2004b; Savard 2009). It was followed by the *Agreement concerning a New Relationship between the Government of Canada and the Cree of Eeyou Istchee* in 2008, which attempted to fix all conflicts concerning the expected responsibilities of the Canadian government in the JBNQA.

Like the JBNQA, the Peace of the Braves was focused on the development of northern Quebec's resources. It included, among other things, an environmental evaluation of the Eastmain-1-A-Sarcelle-Rupert hydroelectric complex and granted more power to the indigenous communities for local management (e.g. the creation of a Cree police force). In addition it comprised the transfer of 4.5 billion dollars in compensation to the Cree over 50 years in exchange for a discontinuation of all proceedings against the Quebec government. Also in its clauses was the establishment of a forest regime adapted to the management of the local fauna and maintaining the environment evaluations for the mining projects detailed in the JBNQA.

Although the role reserved for the Cree in the evaluation of hydroelectric and mining projects as well as for the management of the fauna resources was important under the JBNQA, it was not so for the management of the forest regime (Scott 2005). The Peace of the Braves therefore largely modified the previous conditions related to forest management. Under the JBNQA clearcutting was possible in the southern part of the James Bay territory. This has caused conflicts with Cree families who practiced traditional subsistence activities on their ancestral traplines since the Cree were not consulted or involved in the management of forest operations. This is remedied by including Cree participation into the process in the form of consultations during the various phases of planning and management of the forestry activities (SAA 2002; CCQF 2012).

The Peace of the Braves also sees the creation of new management units composed of three to seven contiguous traplines, the identification of territories of particular interest to the Cree representing up to 1 % of a trapline (e.g. traditional sites, aquatic bird caches, camps) where all forestry activities are prohibited without the express permission of the local Tallyman, the application of specific measures to maintain or improve the habitat of animals important to the Cree (e.g. caribou, beaver, moose) to up to 25 % of each trapline, the maintaining of tree cover on every trapline (e.g. by banning cuts over the entire trapline where 40 % of the area has been cut or burned over the last 20 years), the protection of forests adjacent to rivers and lakes, and the elaboration of a road network in consultation between the beneficiary and the Tallyman to limit the interconnection between traplines and therefore limit the access for poachers (Scott 2005; SAA 2002).

Finally the Peace of the Braves puts in place many medium term measures to conserve the habitats of important animal species. Some of the measures are: only allowing chequerboard cutting in the targeted areas, keeping at least 50 % of the production area of the forests at a height of more than 7 m in which a minimum of 10 % of the area is in forests over 90 years of age, identifying blocks of residual forests to conserve in cooperation with the local Tallymen, and distributing the blocks to favour inter-connexions and limit any interruption in flight cover to a maximum width of 30 m. These measures and others aim to enhance and protect the region's resources by conserving the regions biodiversity and respecting the natural environment (MRNF 2012a).

Finally in 2013 the *Agreement on Governance in the Eeyou Istchee James Bay Territory Between the Crees of Eeyou Istchee and the Government of Quebec* was signed. Its objective was to grant the Cree more autonomy and solve certain conflicts generated by the JBNQA especially concerning category II lands. These conflicts often revolved around the fact that the although the JBNQA stated that the powers and functions of the Government of the Cree Nation would be on equal footing than those of the Baie-James Municipality and the Baie James regional Council, this was not necessarily the case in practice (Gouvernement du Quebec 2011). However none of these conventions or agreements established an independent fourth level of government; rather they recognise the right of the James Bay Cree Nation to manage its territory in cooperation with the federal and provincial governments (Usher 2003; Scott 2005; Salée and Lévesque 2010).

3.1.3 Development Projects

Since the 1960 the James Bay territory has been the focus of a number of development projects. The best known of these is the 1973 hydroelectric dam project on the La Grande River, which flows into the North-East of the James Bay. The La Grande complex was only supposed to be the first phase of a mega project for the James Bay announced on the 30th of April 1971. This mega project was divided into three phases: the La Grande Complex (itself subdivided into three parts), the Grande-Baleine Complex and the Nottaway-Broadback-Rupert Complex (NBR). In its original form, the construction of this structure of complexes included 36 dams, 1000 dykes, the modification of 20 rivers, and the flooding of 23,000 km^2 of land (representing about 2.7 % of the surface area of the Nord-du-Québec AR; Chaplier 2006).

The first two parts of the La Grande Complex consisted of the construction of eight hydro-electric power plants and their reservoirs which covered an area of 13,575 km^2 (Hydro-Québec 2003). The Eastmain, Opinaca and Caniapiscau rivers were diverted to meet the needs of these power plants (Hernandez-Henriquez et al. 2010). These river diversions had significant impacts on the region's fauna and flora, not only because of the large scale flooding that took place but also because of the important changes to the river flow. The main hydrological changes in the James Bay territory were a permanent reduction in flow of approximately 90 % at the mouth of the Eastmain River and a substantial increase in winter flow at the mouth of the La Grande River (Reed et al. 1996; Hernandez-Henriquez et al. 2010). For the Eastmain River, where the Cree community of Eastmain is located, this has not only reduced the river's water levels but has also increased its salinity due to the larger intake of water from the James Bay. These large dams have thus modified the hydrological environment that was used by the Cree to move across the land. It has also resulted in distrust towards waterways by many Cree. This distrust was amplified by the mercury contamination that followed the flooding of land and the decomposition of the submerged vegetation (Kuhnlein and Chan 2000; Desbiens 2007). Although the levels of mercury in the Cree population and in the lakes of the James Bay territory have been closely monitored since the 1980s the misgivings remain to this day (Roebuck 1999).

The second phase of the project, the Grande-Baleine Complex, was met with heavy national and international opposition (Hornig 1999; Carlson 2008). It coincided with a decrease in energy demands and a change of the political party in power in Quebec, which in 1994 lead to the provincial government and Hydro-Québec to suspend it for an indeterminate period of time (Savard 2009). The third phase of the project, the NBR Complex, was permanently shelved at the signing of the Peace of the Braves in 2002 (SAA 2002). The start of works on the modified third part of the La Grande Complex, the Eastmain-1-A-Sarcelle-Rupert entrenched the shelving of the original third phase; as it permanently ends all possible return since the waters from the Rupert River which would be necessary to its completion are diverted towards the Eastmain reservoir (Mulrennan and Scott 2005).

The Eastmain-1-A-Sarcelle-Rupert project includes the creation of the Eastmain reservoir covering an area of 600 km^2, the diverting of 51.7 % of the Rupert River's flow towards the Eastmain reservoir and the La Grande Complex, and the creation of the Eastmain-1A and Sarcelle power stations (Hydro-Québec 2004). As with other phases and parts of the larger project, it also faced opposition, not the least of which was because of the many unknowns linked to the environmental impacts these changes would have on the territory (Savard 2009). The Peace of the Braves allowed the project to go forward following the completion in 2006 of a positive environmental impact assessment. Work on the project started shortly thereafter in January 2007.

The energy created by this massive complex of hydro-electric power stations is distributed throughout Québec by a network of high tension overhead power lines and electric substations (Fig. 3.1). There are six 735 kV lines that cut across the James Bay following a north-south axis sending power from the La Grande complex to various Québec and American cities (Hydro-Québec 2010). The hydro-electric complex also necessitated the construction of a vast transport infrastructure network, including many roads, airports and the Radisson locality. Before the start of this project, no permanent roads existed north of Matagami, with the area's Cree communities being only accessible by plane or boat. The first road completed for the project is the James Bay Road which links Matagami and Radisson over a distance of 620 km (SDBJQ 2011a). It was constructed in 1973 and paved in 1976. This was followed by the construction of Radisson and its associated infrastructure to meet the needs of the future workers. The Transtaïga road starts on km 544 of the James Bay Road and goes until the Caniapiscau reservoir over a distance of 666 km, offering access to the hydroelectric project sites along the La Grande River. At its Eastern limit, the Transtaïga is the furthest northern point in Quebec that can be reached by a permanent road. Both Quebec and Cree outfitter firms specialising in the caribou hunt are located along this road on category III lands. In 1992 the Route du Nord (French for North Road) project is started; which links up Chibougamau to the James Bay Road's 275th km. It is 407 km of gravel road and is mostly used by the lumber industry. Aside from these three main roads (i.e.: James Bay Road, Transtaïga, Route du Nord) smaller roads were created in the 1990s to link up the Cree communities of Chisasibi, Eastmain, Waskaganish and Wemindji to the James Bay Road. These permanent roads and the high tension power lines zigzag over vast distances across the territory and represents clear cuts that can influence fauna and flora distribution and movement in the area (Primack 2006).

The beginning of the 1970s also sees the start of intensive mining exploration in the region. This quickly yields rewards as the area is very rich in minerals such as gold, copper, zinc and silver. Recent explorations have also found deposits of diamonds, lithium and uranium (SDBJQ 2011b; Raymond 2011). The currently active mines are located in the southern part of the James Bay territory whereas the new projects are found throughout the territory. This is partly due to a greater ease of access to the various regions but also to a better understanding of the region's geology. Together the mining and lumber industries form the basis of the region's economy. Timberland account for 81,158 km^2 of the Nord-du-Québec AR and 99 % of these are public lands (CIFQ 2011). Like the mining industry, the lumber industry is also concentrated in the southern part of the James Bay territory, as the smaller size and scarcity of the trees in the treed tundra makes them unsuitable for commercial development (Mulrennan and Scott 2005). Additionally the southern part of the James Bay territory is at the limit of the commercially available forests; above this limit only local community forest management projects are authorised by the provincial government (MRNF 2012b).

In 2005 the Quebec government mentioned for the first time, a potential new northern project called *Plan Nord*. However it was only in 2011 that the plan which originally called for the investment of $80 billion in public and private funds over 25 years is officially presented. It was aimed at the territory located north of the 49° parallel and covered an area of 1,200,000 km^2, encompassing the Eastern James Bay. Its stated aim is to promote the North's mining, energy and tourism potential as well as the social and cultural development of the region, all in a context of respect of the northern population and environment (MRNF 2011). It involved significant investments in communication and infrastructure (e.g. roads, airports). To act as a counterweight to these large scale developments, the plan included aspects of environment and biodiversity protection and social well-being (e.g. housing, social services, health and education). The plan has since been downscaled for its 2015 re-launch and now expects an investment of $50 billion by 2035. Many people have linked the motivations behind this plan to the ones behind the La Grande hydro-complex, which is the government's political will to occupy the north and to integrate it into the national imagery focussed on natural resources development; thus the concern that any positive effects will be outweighed by negative social and environmental impacts remains (Asselin 2011; CNEI 2011; Mulrennan et al. 2012). Additionally, although the plan calls for the respect of existing and future treaties, this might lead to inequalities between First Nations on the Plan Nord's territory as the Cree, Naskapi and Inuit have a convention through the JBNQA while the Innu have none (Leclair 2012). To compound the issue, the Attikamek and Algonquin are not identified by the government as being concerned by this plan as they have no village above the 49th parallel, even though their ancestral lands are partly located there (Asselin 2011). Finally the development of the north means the arrival of a large number of temporary workers coming from southern regions which if not well managed could cause conflicts between indigenous and non-indigenous peoples.

This section illustrates how the majority of anthropic changes in the region happened during the second half of the twentieth century. These development projects had important repercussions on the local ecosystems and through it impacts on daily life of the Cree. The Plan Nord continues on this trend.

3.1.4   The Physical Environment

About 8000 years ago, the Laurentide Ice Sheet covering the James Bay retreated leaving behind the Tyrrell Sea which is the ancestor of today's Hudson Bay. These formations marked the territory. The ice sheet and its retreat are responsible for many of the lakes on the territory while the retreat of the Tyrrell Sea left large expanses of marine clays in valleys, lake beds and flat areas of James and Hudson Bays (Bider 1976; Richard 1979). The retreat of both these land features are still felt in the territory as the area is currently experiencing rapid isostatic rebound, the rate of which is estimated at 0.7 cm/year^{-1} and 1.3 cm/year^{-1} (Andrews 1970; Mitrovica et al. 2000; Lafortune et al. 2006). The region's flatlands are scattered with peatlands and the area is covered in numerous bays and peninsulas as well as many small islands (Richard 1979; Thibault and Payette 2009). The James Bay's central plateau is strewed with many lakes while the eastern part of the territory is hillier. The area's climate is moist continental mid-latitude based on the Koeppen Climate Classification System; which is known for its extreme seasonal variations (-50 °C to $+34$ °C), with warm to cool summers and cold winters (Knight 1968; Laverdière and Guimont 1981). The territory's western border composed of the James and Hudson bays influences the climate by stabilizing precipitations to a rather low average of 765 mm per year (MBJ 2010).

The Boreal forest is found across the James Bay territory. This great coniferous forest is present throughout Eurasia and North America at subarctic latitudes (Hare 1950). In Quebec, it is comprised mainly of black spruce (*Picea mariana*) and tamarack larch (*Larix laricina*), with smaller quantities of white spruce (*Picea glauca*), jack pine (*Pinus banksiana*), paper birch (Betula papyrifera) and white poplar (*Populus tremuloides*), as well as a number of smaller shrubs such as dwarf birch (*Betula nana*), willows (Salix sp.) and a variety of rhododendrons (Rhododendron sp.), laurel (*Kalmia angustifolia*) and berries (e.g.: *Rubus idaeus, Vaccinium sp.*) (Hare 1950; Berkes and Farkas 1990; Arquillière et al. 1990). Towards the north, the forest opens up and gives way to the wooded tundra with its large expanses of lichen interspersed with small trees (Ducruc et al. 1976; Morneau and Payette 1989; Payette et al. 2001; Callaghan et al. 2002). In the wooded tundra these trees rarely exceed 10–15 cm in diameter. On the coast there is a higher concentration of white spruce while the inside of the bays are rich in seagrass (e.g.: *Zostera marina* (eelgrass) and *Carex paleacea*) as well as lichens and ericaceae (Payette and Gagnon 1979; Laverdière and Guimont 1981; Reed et al. 1996).

On the coast and in the lands large expanses of burned areas are common sights as the boreal forest has a short fire cycle varying from 50 to 200 years depending on the region (Payette et al. 1989; Bergeron et al. 2004). In the boreal forest many reviews have shown that the successional processes and flora diversity is controlled primarily through the disturbances due to forest fires. As the forest fire regime in this region is expected to be perturbed through changes in climate, it is foreseeable that the changes will influence the region's vegetation.

The fauna in the James Bay territory is diverse, with 39 separate species identified (MBJ 2010). These species can be divided into two groups: the migratory or mobile species inhabiting a large area and the less mobile or sedentary species inhabiting a smaller area of land. The least mobile species have a tendency to aggregate in areas rich in resources to better meet their needs; this is the case for the majority of the territory's rodent population for example the beaver (*Castor Canadensis*), the musk rat (*Ondatra zibethicus*), the red squirrel (*Tamiasciurus hudsonicus*) and the American hair (*Lepus americanus*) (Bider 1976). The migratory species can survive with a lower concentration of food by constantly moving on the territory; among these species are the caribou (*Rangifer tarandus*), the black bear (*Ursus americanus*), the moose (*Alces alces*), the willow ptarmigan (*Lagopus lagopus*), the Canadian goose (*Branta Canadensis*), and the lesser snow goose (*Chen caerulescens caerulescens*). The James Bay territory also has an important level of fish biodiversity including species such as lake herrings (*Coregonus artedii*), lake whitefish (*Coregonus clupeaformis*), northern pike (*Esox lucius*), capelin (*Mallotus vilosus*), walleye (*Sander vitreus*) and an assortment of trout (*Salvelinus sp.*) and sturgeons (*Acipenseridae sp.*) (Berkes and Mackenzie 1978; Parcs Canada 2011).

Although there is a large variety of species, their abundance is limited by environmental factors and the region's harsh climate. The James Bay territory's environment rests on a careful balance between the climate, the vegetation and the fauna. A change in the climate (e.g.: increase or decrease of rainfall, increase in temperature) could influence the type and number of habitats, the availability and accessibility of nutriments and thus have an impact on the local specie population.

3.2   The Cree of the Eastern James Bay

3.2.1   Eeyou Istchee

The first clues of land occupation in the James Bay territory go back 3500 years however many archaeologists estimate that human presence predates these traces by a further 2000 years (Feit 1995; Morantz 2002; Desbiens 2004a). Before the seventeenth century the Cree were a society comprised of nomadic hunter-gatherers. They moved across a large territory in groups of two to five families subsisting through the outcomes of hunting, fishing, trapping and gathering (Vinette 1996; Peloquin and Berkes 2009). The term 'Eeyou' is used by the Cree of the James Bay territory to identify themselves can be applied to the Nation as a whole as well as to the individual members composing it. With the arrival of Europeans during the seventeenth century, the Cree entered the fur trade; which had already become a vital activity by 1670 (SAA 2009). The Hudson Bay Company and its network of trading posts throughout Northern Canada was a strategic contact point between the Cree and the Europeans, with the presence of the Europeans in the territory not extending much further.

> The company policed its fur trading posts and the English state was involved in protecting the company's links with the outside world, but the native inhabitants of its vast 'properties' maintained indigenous social controls and leadership patterns as the only forms of government in effect at any remove from the posts. (Scott 1989, p. 84)

Neither the Hudson Bay Company nor the federal or provincial governments following confederation seem to have exerted a hold on the administration or management of the Cree lands in the Eastern James Bay; they remained centred on traditional subsistence activities with the fur trade only representing an extension of these activities (Scott 1988; Chaplier 2006). However, a slight social change can be noted with a number of members of the Cree communities starting to settle near the trading posts (Duhaime 2001). During the twentieth century, with the downturn of the fur market, these trading posts were converted into general stores.

The JBNQA brought about a sharp change to the Cree's way of life and forced them to integrate in less than 10 years an unprecedented amount of social, cultural and institutional changes (La Rusic 1979; Desbiens 2004a). Before the conflict caused by the project's announcement, the Cree lived in small isolated communities (or First Nations). These communities rarely had contact with one another as overland transport was complicated and time consuming, air transport was costly and telephone communications were difficult and unreliable (Salisbury 1986). The most common method of transport was by boat however it was limited by climatic conditions and became more difficult the further inland you went.

Following the announcement of the development plan for the James Bay in 1971, all eight of the community chiefs (the ninth community, Oujé-Bougoumou, won't be officially recognised until 2007) met for the first time in Cree history in Mistissini to discuss it. This event marked the start of a tight collaboration between the communities; which was helped along by the fact that even though the communities lived in relative isolation from one another, the James Bay Cree form a historically, socio-economically, culturally and linguistically homogenous group (LaRussic 1979). The cooperation between communities will be further enhanced by the development of the region's infrastructure, which simplified movement and communication.

Through this mobilisation, the Cree gained increased participation in the region's land management and in the creation of managerial organisations (LeBlanc 2009). In addition to the GCC (created to insure Cree interests during the JBNQA negotiations), immediately following the agreement's adoption, the Cree Regional Authority (today replaced by the Cree Nation Government [CNG]) and the Cree Hunters and Trappers Income Security Board are put into place. Other organisations such as the Cree Board of Health and Social Services of James Bay, the Cree School Board, the Cree Trapper's Association to name a few, soon followed between 1975 and 1976. Thus since the announcement of the James Bay development plan, the Cree nation has undergone an important reorganisation and an administrative growth that was unparalleled in the history of any other First Nation group in Canada (Salée and Lévesque 2010). This has created many changes to the available services inside the communities. Today, all Cree communities benefit from the continuous presence of medically trained personnel; however this presence is often secured by one or a few nurses in the smallest villages. In 1971 the Cree population was in the midst of a boom (with more than half of the 5500 Cree being under 20 years of age) caused in part by improvements to food security and medical resources and access (Wills 1984; Salisbury 1986). This boom has continued into 2011. By 2011 14,929 Cree lived in the James Bay and benefited from an increase in life expectancy (Statistique Canada 2012). Schooling has also greatly changed with each of the communities now having at least one primary school on their territory. The expansion of educational and medical services represents advances in the region's socio-economic development by community members and an improvement in living conditions. Like in all things, there is still room for progress (e.g. lack of medical doctors and dentists in some communities, higher than average mortality rate from avoidable or potentially avoidable causes, graduation rate of 23.3 % in 1997); it is however important to note all the advances and changes that were undertaken in such a short time to balance the picture (Larose et al. 2001).

After only two generations following the singing of the JBNQA, more than half of the Cree population had become sedentary and the villages had taken on a size and importance that had yet been unknown in Cree society (LaRusic 1979). Gradually the tipis used by semi-sedentary families gave way to detached or semi-detached houses. Some communities have kept the tipi inside the villages, however not as a living quarters but rather as outdoor kitchens and extra storage. When in the bush, temporary shelters are still used when hunters are too far away to return to their permanent hunting cottages.

Sendarisation and socio-economic changes have had an impact on the region's linguistic aspects. The language spoken by the Cree belongs to the Algonquin linguistic family (SAA 2009). In the Eastern James Bay, there are two major dialects, the coastal and the interior, as well as smaller variations between each community (CCRCCN 1991). The Cree language was an oral one and didn't have a written form until an English missionary living with the Cree of the Western James Bay (in Ontario) adapted in 1840 a syllabic writing originally created for the Ojibwes (Burnaby and Mackenzie 2001). This syllabic writing style was quickly adopted by the Cree and Naskapi of Quebec (Olson and Torrance 1991). It was transmitted not only by the missionaries using religious texts translated into the Cree language but also by the Cree themselves who used it for everyday life such as writing letters and taking stock. At the time of the signing of the agreement, adults and elders (over 40 years of age) rarely spoke another language (La Rusic 1979). The youth from 1950 to 1970 frequented primary and secondary schools in Ontario where they were taught to speak English. Currently, the majority of the population under the age of 70 years old are bilingual, Cree and English. French is also spoken to a lesser extent. The Cree language is one of a few First Nation languages in Canada that has resisted linguistic assimilation due to a high birth rate, the relatively isolated nature of the population and the languages valorisation in school and throughout the community (Goudreau 2003; Collette 2005).

The assimilation that many researchers predicted would be the outcome of these numerous changes did not come about. Instead the Cree Nation of Eeyou Istchee has, like other indigenous communities across Northern Canada, evolved a cultural structure composed of two interrelated spheres of activities, institutions and traditions: the market sphere and the subsistence activity sphere (Usher 2003). The contact between both spheres has thus given a rise to a distinctive and unique Cree culture (Tanner 1979; Feit 1982; Bender and Morris 1991; Petit 2010). Hornig (1999, p. 95) said it best:

> [...] the initial (and ongoing) contact of native cultures with Euro-American or metropol-
> itan societies is seen to produce distinctively new modes of production and social life – new
> formations that are not simply halfway points interpolated between the old and new.

Rather than heading towards a loss of culture the process of assimilating new knowledge into TEK is leading Cree society to towards an integration and reinterpretation of these facts and information.

3.2.2 The Communities

With its nine communities, Eeyou Istchee covers an area of 409,700 km^2. It differs from the James Bay regional administration territory or (Jamésie) because it includes the community of Whapmagoostui and its associated traplines. This community and the land that surrounds it are also integrated into the Kativik regional administration territory of the Nord-du Québec AR for administrative reasons as it is a joint village with the Inuit community of Kuujjuarapilk. Here follows a short description of the nine communities.

3.2.2.1 Chisasibi

Chisasibi is the largest Cree community with 4484 members (Statistics Canada 2012c). It is also the community located furthest north on the eastern cost of the James Bay (Whapmagoostui is located on the Hudson Bay) and is accessible by paved road from the James Bay Road. Chisasibi used to be called Fort George, a name sometimes still used by community members Elders, and was since 1803 the site of a Hudson Bay company trading post. The original village was located on an island close to the current site. It was moved in the 1980 because of the predicted erosion which would ensue following the increase of outflow from the La Grande River due to the hydroelectric complex of the same name.

3.2.2.2 Eastmain

Eastmain is one of the smallest Cree communities with a population of 767 members (Statistics Canada 2012b). It is located on the eastern shore of James Bay at the mouth of the Eastmain River. It is the site of the first permanent Hudson Bay Company trading post, founded in 1723 with the name East Main House. Land access to the community is provided by gravel road from the James Bay Road.

3.2.2.3 Mistissini

Mistissini is an interior community located on the shores of Mistassini Lake and is about 90 km out from the city of Chibougamau. It is the second largest Cree community counting 3427 members (Statistics Canada 2012d). The village was originally the site of various rival trading posts which dealt in the fur trade, such as The Hudson Bay Company's and the North West Company's. It has in the past been called Mistassini as well as Baie du Poste.

3.2.2.4 Nemaska

Nemaska, located in the interior, is the smallest Cree community with 712 members (Statistics Canada 2012e). However it also boasts the head offices of the GCC and the CNG. When the original Hudson Bay Company trading post of Nemiscau (located 60 km north-east of the current village) closed in 1970, the community members dispersed all over the Eastern James Bay's territory. However they were able to finally regroup with the creation in 1980 of the new village of Nemaska.

3.2.2.5   Oujé-Bougoumou

The community of Oujé-Bougoumou and its 725 members are located in the interior on the shores of Opemisca Lake near the village of Chibougamau (Statistics Canada 2012a). It is linked to the Provincial Road 113 by a paved road and houses the Aanischaaukamikw Cree Cultural Institute. Although the community existed before the signing of the JBNQA, its members had been force to move seven times between 1926 and 1970 due to mining prospection (Bosum 2001). The Minister of Indian Affairs had therefore erroneously assigned the members as belonging to other Cree communities such as Mistissini; which explains its exclusion from the JBNQA. Following numerous negotiations a village was build 1992 but it was only formally recognised in 2007 (Otis and Motard 2009).

3.2.2.6   Waskaganish

Waskaganish is the southernmost village located on the eastern shores of the James Bay, and sits at the mouth of Rupert River. It is linked by a gravel road to the James Bay Road. A large community, it counts 2206 members (Statistics Canada 2012f). Site of a Hudson Bay Company trading post established in 1668 by the name of Rupert House, it is still sometimes referred as such by Elder members of the Cree communities.

3.2.2.7   Waswanipi

An interior and southern community, Waswanipi is located along Provincial Road 113. Located near the rivers Chibougamau and Waswanipi, it has 1777 members (Statistics Canada 2012g). The village itself is relatively new having been built in 1978. Up until 1965, the community lived 45 km downstream on Waswanipi River where a Hudson Bay Company trading post was located. That location is now known as Old Post.

3.2.2.8   Wemindji

Located on the eastern shore of the James Bay at the mouth of the Maquatua Rivier, Wemindji has 1378 members (Sayles and Mulrennan 2010; Statistics Canada 2012h). It is accessible by road from the James Bay Road. Before 1958, the community was located at what is now known as Old Factory, an island 25 km south of the current village.

3.2.2.9 Whapmagoostui

The northernmost Cree community, Whapmagoostui is located on the shore of the Hudson Bay at the mouth of Grande-Baleine River. It is part of a conjoint village with the other being the Inuit village of Kuujjuarapik. It has no permanent access roads and all access to the village is done by sea or air. This community is relatively small with 874 members (Statistics Canada 2012i). Like most of the previous communities, it was also the site of a Hudson Bay Company trading post known as Poste de Grande-Baleine and is sometimes referred to as such. Before the creation of the trading post, the area's Cree lived near the Petite-Baleine River.

3.2.3 Traditional Subsistence Activities

In traditional Cree spirituality, it is believed that the *Great Spirit Chishaaminituu* gave them a portion of the land to live on (CTA 2009). Eeyou Istchee is therefore considered their ancestral lands and is the place on which they have based their culture, heritage, spirituality, values, social practices and traditions. The Cree identity is thus strongly linked to their physical environment; with all members of the community having a responsibility to protect and transmit this heritage and by extension of the territory (CTA Committee of Chisasibi 1989; Otis and Motard 2009). Traditional Cree spirituality also believes that animals are sentient beings and that it is the animal that chooses to offers itself to the hunters. Animals will thus have a tendency to offer themselves more often to people they know (i.e. someone who hunts regularly on the land) and who respects them than to a stranger (CTA Committee Chisasibi 1989; Feit 1995).

The subsistence activities practiced on the land (i.e. hunting, trapping, fishing, and gathering) are closely linked to human interrelations, to values, to knowledge and to Cree spirituality (Scott and Feit 1992; Tanner 2007). Consequently subsistence activities extend beyond their simple economic value to incorporate themselves fully into the Cree's identity. As stated by the anthropologist Harvey Feit (1995, p. 102) subsistence activities are "[...] a way of life whose destruction would cause not only an economic and social crisis but a cultural and moral crisis as well." Nowadays, the practice of traditional activities such as hunting, fishing and trapping is also encouraged by the fact that the Cree who practice them regularly can obtain revenue for it through the *Act respecting the Cree Hunters and Trappers Income Security Board* (Feit and Beaulieu 2001; Goudreau 2003). This program, financed by the provincial government, aims to encourage the practice of traditional activities and to stimulate economic activities. Similar programs exist for the Inuit and Naskapi. Because of the continuous valorisation of traditional activities, they remain an important source of food, culture and revenue for the Cree. At least one third of the population continues to have a life style based on subsistence activities and an even larger part of the population divides their time between work in the villages and forest based activities (Delormier and Kuhnlein 1999).

Traditional foods are obtained through subsistence activities, with animals being grouped into six large categories: large mammals (e.g. caribou, moose, bear), fish (e.g. sturgeon, trout, walleye), waterfowl (e.g. Canada goose, snow goose), sea mammals (e.g. polar bear, seal), small mammals (e.g. hare, porcupine), and fur mammals (e.g. beaver, muskrat) (Berkes and Farkas 1978). They also use a number of flora species such as small fruits (e.g. raspberries, blueberries, cloudberry) and medical plants (e.g. black spruce, balsam fir, rhododendrons) (Leduc et al. 2006; Baldea et al. 2010; Uprety et al. 2012). The practice of food sharing allows members to support the ones who cannot hunt be it due to time constraints or health; however over the years with the increased access of Cree society to the outside world, commercially available foods have taken an ever larger place in their daily food habits (Scott 1986; Kuhnlein et al. 2006; Campagna et al. 2011).

Location determines the key species and the timing at which the hunts occur for the communities. Communities located inside the territory (i.e. Mistissini, Nemaska, Oujé-Bougoumou, and Waswanipi) hunt and trap intensively during the winter months; whereas for the coastal communities (i.e. Chisasibi, Eastmain, Waskaganish, Wemindji, and Whapmagoostui) the spring and autumn goose hunts rival in importance the winter hunting season (Scott and Feit 1992). For interior communities moose earns the top space, while geese such as the Canada goose and the Snow goose are the most important source of meat for coastal communities (CREABJNQ 1982). In the northernmost community of Whapmagoostui and some of the northern traplines from Mistissini an important place is reserved for caribou. Other key species are the beaver for southerly communities (i.e. Eastmain, Mistissini, Waswanipi, and Wemindji), in contrast the communities of Chisasibi and Whapmagoostui depend more strongly on fishing (Berkes 1990). The proportion for each species hunted can vary yearly as it relies on a number of factors not least of which is the distribution and abundance of the specie.

The priority of traditional subsistence activities over commercial or sports hunting is guaranteed through various agreements and treaties with the governments (Roué 2009). As stipulated in the JBNQA (SAA 1998, art 24.6.2):

> The principle of priority of Native harvesting shall mean that in conformity with the principle of conservation and where game populations permit, the Native people shall be guaranteed levels of harvesting equal to present levels of harvesting of all species in the Territory.

Additionally there is no system in place to force Cree people to register the kills done for subsistence hunting however restrictions can be imposed when a species is endangered (Gougeon 2010). This mechanism has been put into effect at various intervals such as in 1995 when it was found out that Canada Geese numbers were much lower than previous counts seemed to imply due to a mistake in subspecies identification.

Care must however be taken to avoid perpetuating the western myth of the 'noble savage', branding the First Nations as traditional actors whose only role is to protect nature against the evils of modernisation. This myth can lead researchers to over emphasise or misread the relation between the Cree and their environment and has also been decried by members of the Cree community.

> The romantic 'caretaker' label has been attached to our people by European courts, environmentalists and politicians, and used extensively to dispossess our people of our land. It is unfortunate that many Aboriginal people in Canada have come to believe this myth. We are the owners, not the janitors. (Namagoose 2002, p. 4)

The relationship between the Cree and the land rests on a system of rights and responsibilities which governs land and resource use as well as social relations which was in place long before the arrival of the Europeans (Feit 1995; Desbiens 2004b; CTA 2009). This system does not follow a rigid subordinate hierarchy but rather a mobile rhizome structure where each element can affect another (Desbiens 2007). Subsistence activities are a complex phenomenon that contains a number of subtleties and variations; each Cree community having adapted the system to their needs.

In Cree tradition each of the communities has a territory reserved for their subsistence activities. These territories are then subdivided into family lands call traplines or 'Indoh-hoh Istchee', which were officially recognised in chapter 24 of the JBNQA (Morantz 1978; Tanner 1983; Fraser et al. 2006). These traplines cover the whole of the James Bay territory and vary in size from 350 to 15,000 km^2. It is on these traplines that families practice subsistence activities. In this context, family refers to the extended family group and guests can also be included in this group for various reasons such as the need for hunting partners or a difficulty in accessing their own traplines (Feit 1991). For example, some traplines for the Mistissini community are located along the LaGrande River at the northern limit of the James Bay territory which is without road access. Access to these traplines would either involve time consuming land travel which could be prohibitive if the members are involved in village based work activities or costly plane travel. Also as these divisions predate the development projects, some traplines are partly occupied by these infrastructures limiting the resources available for subsistence activities. Finally some traplines are reserved for use by the whole of the community. Elders play an essential role in this complex and dynamic relation, and are seen as an important source on history, traditions and Cree values; therefore they are consulted when there are conflicts between community members (CTA 2009).

One member from each trapline is selected to ensure that the Eeyou rules are upheld and to assure good management of the territory. Named tallyman in English or 'Kaanoowapmaakin' or 'Indoh-hoh Istchee Ouje-Maaoo' in Cree, he is often chosen by his predecessor because of his competence in the practice of subsistence activities, his knowledge of the environment, the ecosystem and the species that inhabit it, and his knowledge on Cree traditions and rules (Feit 1991). The tallyman is responsible for the management and conservation of the trapline's resources as well as the equitable sharing of the resources amongst the members; however his role is not always practiced in an overt manner (Scott 1986; Berkes 1998). Diplomacy is often key with the tallyman making his views known on a situation in the form of impersonal comments or suggestions and relying on implicit as well as explicit agreements with hunters. Some members have, following the numerous environmental and social changes in Eeyou Istchee started to question the role of the tallyman and have stopped respecting his authority and hunting without permission (CTA 2010). This has led to difficulties in the ecological management of the territory.

Oral traditions called 'Eeyou Weeshou-Wehwun' or 'Eeyou law' are used as a guide and help in the fair and equitable use of the territory and the conservation of the resources. During 2009, the Cree Trappers' Association (CTA), which has as a stated aim to represent Cree hunters in the fur industry and to promote and protect their traditional way of life, undertook to put in writing these oral traditions to ensure their conservation. These customary laws accepted by the Cree population are passed down from one generation to another and can be applied in slightly different manners depending on the community or tallyman. They are based on seven Cree values which are central to their traditional way of life: the courage to do what is considered right, to be honest with themselves and others, to be humble (not boast about your abilities), to show compassion for others, to respect others and the environment, to share knowledge and not be ashamed to ask others for help, and to valorise the wisdom of elders and that which is obtained through experience (Tanner 1979; CTA Committee of Chisasibi 1989; CTA 2010). At the base of these traditions and values is the concept that every member of the Nation has a right to harvest the earth's resources.

The tallyman must undertake inventory or estimations of the availability and state of the resources used for subsistence (e.g. animals, fish, fruit, wood), as well as develop reliable harvesting strategies using harvesting quotas and periods (Berkes 2008). He must also make sure that members respect the limits and rules in place, and that they follow the safe and appropriate use of equipment (e.g. make sure that all the traps are inspected and removed at appropriate times). Finally his role is also to ensure that knowledge, practices and stories linked to the trapline are passed down through the generations.

Although the traplines are divided amongst the families, a tallyman cannot forbid another community member from practicing subsistence activities with the aim of feeding himself. That is to say that hunting, fishing or gathering is allowed to provide immediate sustenance, trapping being not allowed in this case (Scott 1988). Also even though this right exists, any person hunting while traveling on another's land must report to the tallyman the amounts and types of animals killed and, in the case of fur animals, hand over the pelts (CTA 2010). This allows the tallyman to keep a more accurate count of the available resources and avoid any excesses. Gathering is less regulated and all have a right to harvest fruits, mushrooms and non-timber resources (e.g. medicinal plants) for their personal use, although it is preferable to inform the tallyman when doing so.

Through all these customs and rules as well as prolonged contact, tallyman and practicing hunters have actively developed a profound knowledge of the environment and ecosystems. They are therefore likely to notice and in some cases even be on the lookout for changes and departures from the normal way of things.

References

AADNC – Affaires autochtones et Développement du Nord Canada (2009) La Convention de la Baie James et du Nord québécois et la Convention du Nord-Est québécois – Rapport annuel 2005–2006 et 2006–2007. Ministère des Affaires indiennes et du Nord canadien et interlocuteur fédéral auprès des Métis et des Indiens non inscrits, Ottawa

Andrews JT (1970) Present and postglacial rates of uplift for glaciated northern and eastern North America derived from postglacial uplift curves. Can J Earth Sci 7:703–715

Arquillière S, Filion L, Gajewski K, Cloutier C (1990) A dendroecological analysis of eastern larch (Larix laricina) in Subarctic Quebec. Can J For Res 20:1312–1319

Asselin H (2011) Plan Nord. Les autochtones laissés en plan. Rech amérindiennes au Québec 41(1):37–46

Bender B, Morris B (1991) Twenty years of history, evolution and social change in gatherer-hunter studies. In: Ingold T, Riches D, Woodburn J (eds) Hunters and gatherers, volume 1: history, evolution and social change. Berg Publishers Limited, Oxford

Bergeron Y, Gauthier S, Flannigan M, Kafka V (2004) Fire regimes at the transition between mixedwood and coniferous Boreal forest in Northwestern Quebec. Ecology 85(7):1916–1932

Berkes F (1990) Native subsistence fisheries: a synthesis of harvest studies in Canada. Arctic 43 (1):35–42

Berkes F (1998) Indigenous knowledge and resource management systems in the Canadian Subarctic. In: Berkes F, Folke C (eds) Linking social and ecological systems: management practices and social mechanisms for building resilience. Cambridge University Press, Cambridge

Berkes F (2008) Sacred ecology, 2nd edn. Routledge, New York

Berkes F, Farkas CS (1978) Eastern James Bay Cree Indians: changing patterns of wild food use and nutrition. Ecol Food Nutr 7(3):155–172

Berkes F, Mackenzie M (1978) Cree fish names from Eastern James Bay, Quebec. Arctic 31(4):489–495

Bider JR (1976) The distribution and abundance of terrestrial vertebrates of the James and Hudson Bay Regions of Québec. Cah Géogr du Québec 20(50):393–407

Bosum A (2001) Community dispersal and organization: the case of Oujé-Bougoumou. In: Scott C (ed) Aboriginal autonomy and development in Northern Quebec and Labrador. UBC Press, Vancouver

Callaghan TV, Crawford RMM, Eronen M, Hofgaard A, Payette S, Rees GW, Skre O, Sveinbjörnsson B, Vlassova TK, Werkman BR (2002) The dynamics of the Tundra-Taiga boundary: an overview and suggested coordinated and integrated approach to research. Ambio, Spec Rep 12:3–5

Campagan S, Lévesque B, Anassour-Laouan-Sidi E, Côté S, Serhir B, Ward BJ, Libman MD, Drebot MA, Makowski K, Andonova M, Ndao M, Dewailly É (2011) Seroprevalence of 10 zoonotic infection in 2 Canadian Cree communities. Diagn Microbiol Infect Dis 70:191–199

Canobbio É (2009) Géopolitique d'une ambition inuite. Le Québec face à son destin nordique. Septentrion, Québec

Carlson HM (2008) Home is the hunter: the James Bay Cree and their land. UBC Press, Vancouver

CCQF – Conseil Cris-Québec sur la foresterie (consulted 2012) Mission/mandat. www.ccqfcqfb. ca/fr/0101_mission_mandat.html

CCRCCN – Comité chargé du réexamen de la commission Crie-Naskapie (1991) Report of the inquiry into the Cree-Naskapi Commission. Canada

Chaplier M (2006) Le conflit à la baie James: Pour une anthropologie de la nature dans un contexte dynamique. Civilisations 55:103–115

CIFQ – Conseil de l'industrie forestière du Québec (2011) Portraits forestiers régionaux: 10- Nord-du-Québec, Québec: 2011, Quebec. www.cifq.qc.ca/html/francais/centre_mediatique/portrait_10.php

CNEI – Cree Nations of Eeyou Istchee (2011) Cree vision of plan nord. Canada

Collette V (2005) Rétention linguistique et changement social à Mistissini. Etudes Inuit Stud 29 (1–2):207–219

CRBJ – Conférence régionale de la Baie-James (2011) Portrait de la Jamésie. www.crebj.ca

CREABJNQ – Comité de recherche sur l'exploitation par les autochtones de la Baie-James et du Nord québécois (1982) Terre d'abondance. Étude sur l'exploitation de la faune par les Cris de la Baie-James, de 1972–1979. Québec

CTA – Cree Trappers' Association (2009) Traditional Eeyou hunting law. Eastmain Canada, unpublished

CTA – Cree Trappers' Association (July 2010) Traditional Eeyou hunting law: frequently asked questions. Eastmain Canada, unpublished

CTACC – Cree Trappers Association's Committee of Chisasibi (1989) Cree Trappers speak. James Bay Cree Cultural Education Center, Québec

CWGPN – Cree Working Group on the Plan Nord (2011) Cree vision of Plan Nord. Grand Council of the Cree/Cree Regional Authority, Nemaska

Delormier T, Kuhnlein HV (1999) Dietary characteristics of Eastern James Bay Cree women. Arctic 52(2):182–187

Desbiens C (2004a) Producing North and South: a political geography of hydro development in Québec. Can Geogr 48(2):101–118

Desbiens C (2004b) Nation to nation: defining new structures of development in Northern Quebec. Econ Geogr 80(4):351–366

Desbiens C (2007) 'Water all around, you cannot even drink': the scaling of water in James Bay/Eeyou Istchee. Area 39(3):259–267

Ducruc J-P, Zarnovican R, Gerardin V, Jurdant M (1976) Les régions écologiques du territoire de la baie de James: caractéristiques dominantes de leur couvert végétal. Cah Géogr du Québec 20 (50):365–392

Duhaime G (dir) (2001) Le Nord: habitants et mutations. Les Presses de l'Université Laval, Québec

Feit HA (1982) The future of hunters within nation-states: anthropology and the James Bay Cree. In: Leacock EB, Lee R (eds) Politics and history in band societies. Cambridge University Press, New York

Feit HA (1991) Gifts of the land: hunting territories, guaranteed incomes and the construction of social relations in james Bay Cree society. Senri Ethnol Stud 30:223–268

Feit HA (1995) Hunting and the quest for power: the James Bay Cree and Whitemen in the 20th century. In: Morrison BR, Wilson RC (eds) Native peoples: the Canadian experience, 2nd edn. McCelland & Stewart Publishers, Toronto

Feit HA, Beaulieu R (2001) Voices from a disappearing forest: government, corporate, and Cree participatory forestry management practices. In: Scott C (ed) Aboriginal autonomy and development in Northern Quebec and Labrador. UBC Press, Vancouver

Fraser DJ, Coon T, Prince MR, Dion R, Bernatchez L (2006) Integrating traditional and evolutionary knowledge in biodiversity conservation: a population level case study. Ecol Soc 11(2): art4

Froschauer K (1999) White gold: hydroelectric power in Canada. UBC Press, Vancouver

Goudreau É (2003) Les autochtones et le Québec. In: Weidmann-Koop M-C (ed) Le Québec aujourd'hui: identité, société et culture. Les Presses de l'Université Laval, Québec

Gougeon N (2010) Co-management of the migratory caribou herds in Northern Québec: the perspective of the hunting, fishing and trapping coordinating committee. Rangifer, Special Issue (20):39–45

Gouvernement du Québec (27 mai 2011) Accord-Cadre entre les Cris d'Eeyou Istchee et le gouvernement du Québec sur la gouvernance dans le territoire d'Eeyou Istchee Baie-James. Québec

Hamelin L-E (1998) L'entièreté du Québec: le cas du Nord. Cah Géogr du Québec 42 (115):95–110

Hare KF (1950) Climate and zonal divisions of the Boreal forest formation in Eastern Canada. Geogr Rev 40(4):615–635

Hernandez-Henriquez MA, Mlynowski TJ, Dery SJ (2010) Reconstructing the natural streamflow of a regulated river: a case study of La Grande Rivière, Quebec, Canada (Case Study). Can Water Res J 35(3):301–316

Hogue C, Bolduc A, Larouche D (1979) Québec, un siècle d'électricité. Libre Expression, Montréal

Hornig JF (ed) (1999) Social and environmental impacts of the James Bay hydroelectric project. McGill-Queen's University Press, Quebec

Hydro-Québec (2003) La Grande Hydroelectric Complex: fish communities. Quebec

Hydro-Québec Production (2004) Centrale de l'Eastmain-1-A et dérivation Rupert: Étude d'impact sur l'environnement – Rapport de synthèse, Québec

Hydro-Québec (2010) Rapport Annuel 2010: grands équipements, Québec Canada

Knight R (1968) Ecological factors in changing economy and social organization among the Rupert House Cree. Anthropology Papers, National Museum of Canada, Department of the Secretary of State, Ottawa Canada

Kuhnlein HV, Chan HM (2000) Environment and contaminants in traditional food systems of northern indigenous peoples. Annu Rev Nutr 20:595–626

Kuhnlein H, Erasmus B, Creed-Knashiro H, Englberger L, Okeke C, Turner N, Allen L, Bhattacharjee L (2006) Indigenous peoples' food systems for health: finding interventions that work. Public Health Nutr 9(8):1013–1019

La Rusic IE (dir) (1979) La négociation d'un mode de vie: la structure administrative découlant de la Convention de la Baie James: l'expérience initiale des Cris. ssDcc Inc, Montréal

Lafortune V, Filion L, Hétu B (2006) Impacts of Holocene climatic variations on alluvial fan activity below snowpatches in Subarctic Québec. Geomorphology 76:375–391

Larose F, Bourque J, Terrisse B, Kurtness J (2001) La résilience scolaire comme indice d'acculturation chez les autochtones: bilan de recherches en milieux innus. Rev Sci de l'éducation 27(1):151–180

Laverdière C, Guimont P (1981) Géographie physique de la Grande Ile, Littoral québécois de la Mer d'Hudson. Société de développement de la Baie James. Aménagement Régional, Québec

Le Blanc K (2009) Évaluation de la participation des Cris dans la procédure d'évaluation environnementale de la Convention de la Baie James et du Nord québécois. Mémoire présenté à la Faculté des études supérieures en vue de l'obtention du grade de Maîtres ès sciences (M. Sc) en géographie. Département de géographie, Faculté des Arts et des Sciences Université de Montréal, Canada

Leclair J (2012) L'effet structurant des droits reconnus aux peuples autochtones sur le débat entourant le Plan Nord. Soc Sci Res Netw. http://dx.doi.org/10.2139/ssrn.2046528

Leduc C, Coonishish J, Haddad P, Cuerrier A (2006) Plants used by the Cree nation of Eeyou Istchee (Quebec, Canada) for the treatment of diabetes: a novel approach in quantitative ethnobotany. J Ethnopharmacol 105:55–63

MAMOT – Ministère des Affaires municipales et de l'Occupation du territoire (2014) L'organisation municipale et régionale au Québec en 2014. Gouvernement du Québec, MAMOT, Québec

MAMROT – Ministère des Affaires municipales régions et occupation du territoire du Québec (2011) Baie-James, 2011. Gouvernement du Québec, Québec. www.mamrot.gouv.qc.ca

MBJ – Municipalité de Baie-James (2010) Territoire de la Baie-James, 2010. Québec. http://municipalite.baie-james.qc.ca/html/territoire_bj.php

MBJ – Municipalité de Baie-James (2011) Localités, 2010. Québec. www.municipalite.baiejames.qc.ca

MDDEPQ – Ministère du Développement durable, Environnement et Parcs du Québec (consulted 2011a) Région administrative du Nord-du-Québec, Portrait socio-économique de la région, 2002. Quebec. www.mddep.gouv.qc.ca

MDDEPQ – Ministère du Développement durable, Environnement et Parcs du Québec (consulted 2011b) Évaluation environnementale de projets en milieu nordique, 2002. Quebec. www. mddep.gouv.qc.ca

MDECRQ – Ministère du Développement économique Canada pour les régions du Québec (consulted 2011) Nord-du-Québec (10) – Profil socioéconomique, 2010-11-15. Quebec. www.dec-ced.gc.ca

Mercier G, Ritchot G (1997) La Baie James. Les dessous d'une rencontre que la bureaucratie n'avait pas prévue. Cah Géogr du Québec 41(113):137–169

Minister of Justice of Canada (Last amended 2015) Indian Act R.S.C. 1985, c. I-5, art.18(1)

Mitrovica JX, Forte AM, Simons M (2000) A reappraisal of postglacial decay times from Richmond Gulf and James Bay, Canada. Geophys J Int 142:783–800

Morantz T (1978) The probability of family hunting territories in eighteenth century James Bay: old evidence newly presented. In: Cowan W (ed) Proceedings of the Ninth Algonquian conference. Carleton University, Ottawa

Morantz T (2002) The White Man's Gonna Getcha: the colonial challenge to the Crees in Quebec. McGill-Queen's University Press, Montreal

Morneau C, Payette S (1989) Postfire lichen – spruce woodland recovery at the limit of the Boreal forest in Northern Quebec. Can J Bot 67:2770–2782

MRNF – Ministère des Ressources naturelles et de la Faune du Québec (2011) Plan Nord. Faire le Nord ensemble. Le chantier d'une génération. Québec

MRNF – Ministère des Ressources naturelles et de la Faune du Québec (consulted 2012a), La Loi sur les forêts. Quebec. www.mrnf.gouv.qc.ca/forets/quebec/quebec-regime-gestion-loi.jsp

MRNF – Ministère des Ressources naturelles et de la Faune du Québec (consulted 2012b) Limite nordique des forêts attribuables pour un aménagement forestier durable. Quebec. www.mrnf. gouv.qc.ca/forets/amenagement/amenagement-limite-nordique.jsp

Mulrennan ME, Scott CH (2005) Co-management – an attainable partnership? Two cases from James Bay, Northern Quebec and Torres Strait, Northern Queensland. Anthropologica 47(2):197–213

Mulrennan ME, Marc R, Scott CH (2012) Revamping community-based conservation through participatory research. Can Geogr 56(2):243–259

Namagoose B (2002) A message from the newsletter editor, Eeyou Eenou nation: the voice of the people. Embassy of the Cree Nation, Ottawa

Nistor Baldea LA, Martineau LC, Benhaddou-Andaloussi A, Arnason JT, Lévy É, Haddad PS (2010) Inhibition of intestinal glucose absorption by anti-diabetic medicinal plants derived from the James Bay Cree traditional pharmacopeia. J Ethnopharmacol 132:473–482

Olson DR, Torrance N (eds) (1991) Literacy and orality. Cambridge University Press, New York

Otis G, Motard G (2009) De Westphalie à Waswanipi: la personnalité des lois dans la nouvelle gouvernance crie. Cah de droit 50(1):121–152

Parcs Canada (consulted 2011) Aires marines nationales de conservation du Canada, Baie James, novembre 2006. Canada. http://www.pc.gc.ca/progs/amnc-nmca/systemplan/itm1-/arc10_f. asp

Payette S, Gagnon R (1979) Tree-line dynamics in Ungava Peninsula, Northern Quebec. Holarct Ecol 2:239–248

Payette S, Morneau C, Sirois L, Desponts M (1989) Recent fire history of the Northern Québec biomes. Ecology 70(3):656–673

Payette S, Fortin M-J, Gamache I (2001) The subarctic forest-tundra: the structure of a biome in a changing climate. Bioscience 51(9):709–718

Peloquin C, Berkes F (2009) Local knowledge, subsistence harvests and social-ecological complexity in James Bay. Hum Ecol 37:533–545

Petit J-G (2010) Cris et Inuit du nord du Québec: deux peuples entre tradition et modernité (1975-2010). In: Petit J-G, Bonnier VY, Aatami P, Iserhoff A (dir) Les Inuit et les Cris du nord du Québec. Presses universitaires de Rennes, Quebec

Primarck RB (2006) Essentials of conservation biology, 4th edn. Sinauer Associates, Sunderland

Raymond D (2011) Le lithium au Québec: les projets miniers d'actualité, Ministère des Ressources naturelles et de la Faune, Québec: juin 2011. Gouvernement du Québec, Québec. www.mrnf.gouv.qc.ca/mines/quebec-mines/2011-06/lithium.asp

Reed A, Benoit R, Michel J, Lalumière R (1996) Utilisation des habitats côtiers du nord-est de la baie James par les bernaches. Publication hors-série No.92, Service canadien de la faune, Canada

Richard P (1979) Contribution a' l'histoire postglaciaire de la végétation au nord-est de la Jamésie, Nouveau-Québec. Géog Phys Quatern 33(1):93–112

Roebuck BD (1999) Elevated mercury in fish as a result of the James Bay hydroelectric development: perception and reality. In: Hornig JF (ed) Social and environmental impacts of the James Bay hydroelectric project. McGill-Queen's University Press, Montreal

Roué Marie (2009) Une oie qui traverse les frontières. La bernache du Canada. Ethnologie française 2009/1, Tome XXXIX

Royer M-JS (2012) L'interaction entre les savoirs écologiques traditionnels et les changements climatiques: les Cris de la Baie-James, la bernache du Canada et le caribou des bois. Thèse présentée à la Faculté des Arts et Sciences en vue de l'obtention du grade de Philosophiae Doctor (PhD) en géographie, Université de Montréal, Montréal Canada

SAA – Secrétariat aux Affaires autochtones (1998) James Bay and Northern Québec agreement and complementary agreements, 1998th edn. Les Publications du Québec, Québec

SAA – Secrétariat aux affaires autochtones (2002) Entente concernant une nouvelle relation entre le Gouvernement du Québec et les Cris du Québec, Québec Canada (3.1.c)

SAA – Secrétariat aux affaires autochtones (2009) Profil des Nations: Cris, Québec: 19 mai 2009. Gouvernement du Québec, Québec. www.saa.gouv.qc.ca/relations_autochtones/profils_nations/cris.htm

Salée D, Lévesque C (2010) Representing Aboriginal self-government and first nations/state relations: political agency and the management of the Boreal forest in Eeyou Istchee. Int J Can Stud 41:99–135

Salisbury RF (1986) A homeland for the Cree: regional development in James Bay, 1971–1981. McGill-Queen's University Press, Montreal

Savard S (2009) Les communautés autochtones du Québec et le développement hydroélectrique: un rapport de force avec l'État, de 1944 à aujourd'hui. Rech amérindiennes au Québec 39 (1–2):47–60

Sayles JS, Mulrennan ME (2010) Securing a future: Cree hunters' resistance and flexibility to environmental changes, Wemindji, James Bay. Ecology and Society 15(4):Art.22. www.ecologyandsociety.org/vol15/iss4/art22/

Scott C (1986) Hunting territories, hunting bosses and communal production among coastal James Bay Cree. Anthropologica 28(1/2):163–173

Scott C (1988) Property, practice and aboriginal rights among Quebec Cree hunters. In: Ingold T, Riches D, Voodburn J (eds) Hunters and gatherers, volume 2: property, power and ideology. Berg Publishers Limited, Oxford

Scott C (1989) Ideology of reciprocity between the James Bay Cree and the Whiteman State. In: Skalník P (ed) Outwitting the state, political anthropology, vol 7. Transaction Publishers, New Brunswick

Scott C (2005) Co-management and the politics of Aboriginal consent to resource development: the agreement concerning a new relationship between le Gouvernement du Québec and the Cree of Québec. In: Murphy M (ed) Canada: the state of the federation 2003: reconfiguring aboriginal-state relations. McGill-Queen's University Press, Montreal

Scott CH, Feit HA (1992) Income security for Cree hunters: ecological, social and economic effects, Monograph series. McGill Program in the Anthropology of Development, Montreal

SDBJQ – Société de développement de la Baie-James du Québec (consulted 2011a) Historique, Gouvernement du Québec: 2008–2009. Quebec Canada. www.sdbj.gouv.qc.ca/fr/societte/historique/

SDBJQ – Société de développement de la Baie-James du Québec (consulted 2011b) Les projets régionaux: Projets miniers, Gouvernement du Québec, Québec: 2008–2009. Quebec. www. sdbj.gouv.qc.ca/fr/projets_developpement/projets_miniers/

Statistics Canada (2012a) Oujé-Bougoumou, Québec (Code 2499818) and Nord-du-Québec, Quebec (Code 2499) (table). Census profile, 2011 census. Statistics Canada Catalogue no.98-316-XWE. Ottawa

Statistics Canada (2012b) Eastmain, Québec (Code 2499810) and Nord-du-Québec, Quebec (Code 2499) (table). Census profile, 2011 census. Statistics Canada Catalogue no.98-316-XWE. Ottawa

Statistics Canada (2012c) Chisasibi, Québec (Code 2499814) and Nord-du-Québec, Quebec (Code 2499) (table). Census profile, 2011 census. Statistics Canada Catalogue no.98-316-XWE. Ottawa

Statistics Canada (2012d) Mistissini, Québec (Code 2499804) and Nord-du-Québec, Quebec (Code 2499) (table). Census profile, 2011 census. Statistics Canada Catalogue no.98-316-XWE. Ottawa

Statistics Canada (2012e) Nemaska, Québec (Code 2499808) and Nord-du-Québec, Quebec (Code 2499) (table). Census profile, 2011 census. Statistics Canada Catalogue no.98-316-XWE. Ottawa

Statistics Canada (2012f) Waskaganish, Québec (Code 2499806) and Nord-du-Québec, Quebec (Code 2499) (table). Census profile, 2011 census. Statistics Canada Catalogue no.98-316-XWE. Ottawa

Statistics Canada (2012g) Waswanipi, Québec (Code 2499802) and Nord-du-Québec, Quebec (Code 2499) (table). Census profile, 2011 census. Statistics Canada Catalogue no.98-316-XWE. Ottawa

Statistics Canada (2012h) Wemindji, Québec (Code 2499812) and Nord-du-Québec, Quebec (Code 2499) (table). Census profile, 2011 census. Statistics Canada Catalogue no.98-316-XWE. Ottawa

Statistics Canada (2012i) Whapmagoostui, Québec (Code 2499816) and Nord-du-Québec, Quebec (Code 2499) (table). Census profile, 2011 census. Statistics Canada Catalogue no.98-316-XWE. Ottawa

Statistique Canada (consulted 2012) Profil pour le Canada, les provinces, les territoires, les divisions de recensement et les subdivisions de recensement, Recensement de 2006: Nord-du-Québec. Canada. www12.statcan.ca/census-recensement/2006/index-fra.cfm

Swyngedouw E (2007) Technonatural revolutions: the scalar politics of Franco's hydro-social dream for Spain, 1939–1975. Trans Inst Br Geogr 32(1):9–28

Tanner A (1979) Bringing home animals: religious ideology and mode of production of the Mistassini Cree Hunter, Institute of Social and Economic Research. Memorial University of Newfoundland, Canada

Tanner A (1983) Algonquin land tenure and state structures in the North. Can J Nativ Stud 3(2):311–320

Tanner A (2007) The nature of Quebec Cree animist practices and beliefs. In: Laugrand FB, Oosten JG (dir) La nature des esprits dans les cosmologies autochtones. Les Presses de l'Université de Laval, Quebec

Thibault S, Payette S (2009) Recent permafrost degradation in bogs of the James Bay area, Northern Quebec, Canada. Permafr Periglac Process 20:383–389

Transport Québec (consulted 2011) Nord-du-Québec: Territoire et population, Gouvernement du Québec, Québec: 2007. Quebec. www.mtq.gouv.qc.ca/portal/page/portal/ministere/ministere/plans_transport/nord_quebec/territoire_population

Uprety Y, Asselin H, Dhakal A, Julien N (2012) Traditional use of medicinal plants in the Boreal Forest of Canada: review and perspectives. J Ethnobiol Ethnomed 8(1):Art.7. www. ethnobiomed.com/content/8/1/7

Usher PJ (2003) Environment, race and nation reconsidered: reflections on aboriginal land claims in Canada. Can Geogr 47(4):365–382

Vincent S (1992) La révélation d'une force politique: les Autochtones. In: Daigle G, Rocher G (dir) Le Québec en jeu: comprendre les grands défis. Presses de l'Université de Montréal, Montréal

Vinette D (1996) École, parents amérindiens et changements sociaux: la perception d'un intervenant non autochtone. Lien Social et Politiques 35:23–35

Washaw Sibi Eeyou (consulted 2011) Home. www.washawsibi.ca

White R (1995) The organic machine: the remaking of the Columbia river. Hill and Wang, New York

Wills RH (1984) Conflicting perceptions: Western economics and the Great Whale River Cree. Tutorial Press, Canada

Chapter 4
Traditional Subsistence Activities and Change

4.1 Methodology

4.1.1 A Case Study in Ethnoecology

Undertaking a case study was a clear and deliberate choice as it allows to identify the potential effects of climate change in a holistic way by retaining as many variables as possible and thus placing the changes into a real-life context while still allowing for the possibility to extrapolate the results (Yin 2003; Stake 2005; Baxter and Jack 2008). It is for this reason that placing the subject matter into context (e.g. social, cultural, historical, geographical, political, and environmental) is crucial and that the first part of this book was dedicated to that process.

Ethnoecology rests at the confluence of natural and social sciences. This approach is often used to study human societies in their local setting with particular emphasis placed on their knowledge and their use of the environment's resources and biodiversity (Conklin 1954; Toledo 2002). Ethnoecology bases its approach on the study of a bi-directional relationship between humans and the environment taking into account both how humans use their environment (and the tools that they develop for this exercise) as well as the way the environment shapes the development of human society through knowledge development, perceptions and practices (Godelier 1984; Bahuchet 1991; Toledo 2002). These interrelations occur through social structures such as local knowledge and representation of the environment, social values, cosmogony, and the local cultural identity; which are used for the management and use of the local landscape (Johnson 2010).

An ethnoecological study pays attention to the context in which the studied group is located and must systematically integrate an analysis of precise behaviours. The researcher must then take into account context and behaviour to understand how they play a role in knowledge development (Milton 1997; Brunois 2005). This is important as it provides the researcher with a better ability to understand the intricacies linked to a group which is not his own.

© The Author(s) 2016
M.-J.S. Royer, *Climate, Environment and Cree Observations*,
SpringerBriefs in Climate Studies, DOI 10.1007/978-3-319-25181-3_4

Ethnoecology rests at the junction of landscape's nature, production and culture dimensions, and is thus a privileged approach when studying TEK (Toledo 1992; Ellen 2006). Although all cultures are capable of grasping objective elements of nature, the way these elements are perceived and identified are specific to the culture. Nature and culture are thus multi-faceted without the possibility of a single description (MacCormack and Strathern 1980; Cronon 1996). Studies undertaken by Watanabe have reinforced the importance of the nature-culture relation showing that even in the case of animals the objective environment is not enough to explain their specific behaviour; their behaviour being influenced by their subjective perception of the environment (Watanabe 1964). The perception of nature can also vary inside of groups based on the position occupied by individual (Watanabe 1983). This study therefore targeted adult Cree hunters to create a homogenous sampling group even though other groups (e.g. youths, non-hunters [such as a larger proportion of elders and women]) could also have brought different perceptions and observations (Haraway 1988).

4.1.2 Data Sources and Collection

The information detailed in this book was collected during a larger study on climate change and modification of species distribution in the James Bay territory. The qualitative data documented further in the text is based on field research conducted in 2010. This study was limited to adult Cree Trappers' Association (CTA) members who at the time of the research represented 73.5 % of the adult population (CTA 2010d). Junior members (under 18 years of age) were excluded as they would not be able to recall the weather over a long enough time period. Questions asked of participants required them to compare the current situation to when they were 16 years old.

The goal of the fieldwork was to obtain data on the observations of Cree hunters on the effects of climate change on annual environmental conditions. Four villages were targeted: Mistissini, Nemaska, Eastmain and Waswanipi. Nemaska and Waswanipi were selected as they were hosting the general assemblies for the Grand Council of the Cree and Cree Regional Administration (GCC-CRA), and the CTA respectively. Sampling participants at these reunions gave access to CTA members from all the nine communities. This allowed the additional sampling returns of participants from Chisasibi, Waskaganish, Wemindji and Whapmagoostui. Our participation at these assemblies gave us further insight into the management of traditional subsistence activities in Eeyou Istchee. Mistissini was selected as it represented the eastern most village in the James Bay territory while Eastmain is at the western limit. This combination also allowed for two villages in proximity to non-indigenous communities (Mistissini and Waswanipi) and two more isolated (Nemaska and Eastmain).

While all researchers would like to think that everything will always go according to plan that is unfortunately rarely the case. Maybe it is not so unfortunate as many happy discoveries have happened when things deviated from the expected. However to avoid the unpleasant aspects of the unexpected, planning and a solid understanding of the subject matter is important thus allowing to obtain pertinent information in the most systematic manner possible (Gumuchian and Marois 2000).

Any fieldwork carried out in Quebec is explicitly or implicitly undertaken in a bilingual context as strategic information may be provided in either French or English. With the fieldwork being in Eeyou Istchee an extra level was added as the Cree language is extensively used in their daily life. Thanks to active and involved members of the CTA, messages where translated into Cree when necessary; however most communications (i.e. interviews, questionnaires) with Cree participants were in English. It was also necessary to familiarise ourselves with the specific terminology used for traditional subsistence activities and in Cree TEK (as we detailed in earlier chapters) to ensure a clearer understanding of the responses. The importance of language cannot be understated as it often plays a significant role in the way the environment is conceived and interpreted (Mülhäusler 1995; Sutherland 2003; Maffi 2005).

It is also necessary to take into account the season as the time of year plays a role in people's mobility. In summer the Cree often travel between villages and traplines to attend various activities and meetings. This rendered a probabilistic sampling method more problematic. This as well as the fact that we were targeting adult hunters who had a similar socio-economic, linguistic and ethnic profile it was decided that we could go forward with a non-probabilistic sampling method. Two types of data collection methods were used: questionnaires and interviews.

4.1.3 Questionnaires

Questionnaires were used as they allowed the collection of precise information on the population's local observations (Huntington 2000b). The majority of questions were close-ended asking participants to select an answer among a gradient, but it also included in strategic places open-ended questions to allow them to detail their answers. As previously mentioned the information presented here was part of a larger study. The questionnaire was divided into six sections. The first section included base information on the participants (i.e. age, gender, community) which were used for classification and analysis. The second section was questions about the climate (e.g. snowfall, rainfall, ground snow accumulation, ice conditions, temperature). Sections three, four and five were linked to hunting and animal behaviour. Although their results are not included here, elements from the answers have been included in the way of insights into Cree traditional hunting and observation methods. Finally the sixth section was about intergenerational transmission of knowledge. Again like the three previous sections they are not explicitly detailed here but were useful in allowing a deeper understanding of the context associated with the answers given in section two.

Questionnaire participants were selected using blind-sample questionnaires. All were selected from public spaces in the four targeted villages. One hundred questionnaires were distributed and of these 30 were returned completed. This represents a rather small sample size however because of the type of research as well as the ethnoecological approach used, the sample size was enough to allow the identification of large trends in the results (Geertz 1973; Sandelowski 1995; Williams 2000; Small 2009).

Questionnaires were distributed between men and women however most women refused to answer preferring to pass them on to male members of their families. Thus of the returned questionnaires 27 were completed by men, 2 by women and 1 had no information related to gender. This refusal of female members of the community to answer was partly linked to gender roles. It is traditionally men who hunt while women prepare the kills. As the questionnaire was identified with hunting activities women members did not feel that they were apt to answer, while the two women who answered were active hunters. Men who didn't actively hunt also refused to answer, with one member stating: "I can't answer that, I am too old to hunt. I don't know anymore." (Anonymous 2010c) The limited return of questionnaires by female members does not permit the identification of potential gender variations in observations. At the time of the sampling, participants were between 30 and 69 years of age. Twenty-one participants originated from coastal communities and nine from inland communities.

The questionnaire data was analysed a first time all communities together to look for trends based on participants' age group. The participants were divided into age groups of 10 years. We had nine by participants aged 30–39, eight by 40–49 year olds, nine by 50–59 year olds and four over the age of 60. The questions asked participants to compare the current situation with the one when they were 16 years of age. As the current age of participants was a known variable, it was then simple to work back to identify the reference period. Sixteen years of age was selected as it gave a precise comparison point which would be easily usable in the analysis as well as because memories at 16 years of age are often more precise then those from a younger age as they are often associated with key life periods that facilitate their recollection (Czernochowski et al. 2005; Ghetti and Angellini 2008).

Trend analysis was done on a coastal and inland location of communities. However no north-south comparison was undertaken as traplines for each community can extend quite far to the north (e.g. some Mistissini traplines are located along the Transtaïga). Therefore any north-south analysis done using the location of the participant's village ran the risk of returning false results. Trend analysis was thus limited to coastal-inland and age groups.

4.1.4 Interviews

Interviews were used to complete and enrich the information obtained by the questionnaires. They were of two types. The first was in-depth semi-structured interviews where the three respondents (one each in Eastmain, Mistissini and Nemaska) were selected using a snowball sampling technique (Biernacki and Waldorf 1981). The participants identified for these questionnaires were perceived by CTA members as having extensive knowledge of the local environment and climatic conditions. The other was short open-ended interviews with 18 community members selected using an accidental sampling method. Ten of those interviews were in the coastal community of Eastmain while eight were in inland communities (i.e. Mistissini, Nemaska and Waswanipi).

The questions were grouped around the studie's research theme but participants were free to direct the conversation towards aspects of climate change impacts that interested them the most. Participants were encourages to comment and describe their experiences relating to the impacts of climate change. The comments were varied from "there are less frogs" to "It is getting harder to predict weather" (Anonymous 2010c). They showcased the diversity of preoccupations that the Cree hunters have on the impacts of climate change on the local environment.

Although interviews are subjective in nature, they can be useful when used in conjunction with other data collection methods. Additionally they were able to bring to light sociocultural and environmental problems pertinent to Cree community members. An inductive thematic analysis was done on the information gathered during the interviews. The interviews were decontextualized to extract independent information. These were then grouped based on a predefined list of themes. Finally the extracts were contextualised again to make them intelligible and meaningful (Paillé 1996).

4.1.5 Data Triangulation

Data triangulation which stresses the need for a plurality in data collection methods helps verify the repeatability of an observation or interpretation (Flick 2002; Stake 2005). This study used a three step analytic framework. In the first step, the data from questionnaires and interviews was examined separately. Data from meteorological stations was also analysed separately by Royer et al. (2013). Questionnaire and interview data was cross-tabulated by age group and community location.

The second step combined the qualitative data from questionnaires, interviews, and CTA community workshops along key themes. The community workshops were conducted during the same time period as our study by the CTA on similar themes. It was decided that this study would avoid duplicating their work and thus eliminate redundancy in research and study fatigue among participants.

Third, the qualitative data was compared and combined to the quantitative meteorological station data. This offered new perspectives on the impacts of climate change and emphasised aspects that were not immediately perceived by using the quantitative or qualitative data alone.

4.2   The Climate and the Environment

4.2.1   Available Weather Stations

The territory of the James Bay territory has had a number of weather stations from various organisations such as Environment Canada (starting in 1920), le Ministère du Dévelopement durable, de l'Environnement et des Parcs du Québec (intermittently between 1970 and 1995) and Hydro-Québec (since 1980). However the majority of these stations present a notable variability in their data because of the different standards that were in use and because of missing data sets (Meunier 2007; Brown 2010). Thus only three stations were available with an almost complete data set covering the time period of 1974–2000 (i.e. Chapais 2: 1974–1998 [located 49°47′N, 74°51′W at 396 m altitude], Kuujjuarapik A: 1978–2002 [located 55°17′N, 77°45′W at 10 m altitude], La Grande Rivière A: 1978–2002 [located 53°38′N, 77°42′W at 195 m altitude]). This low density of stations with long-term measurements for such a large territory is problematic as it causes difficulties in distinguishing trends between the areas (e.g. coastal vs inland, North vs South), and increases the potential error margin in trend analysis (Jones 1997). Out of territory stations were not able to complete the data because of their limited data coverage or locations did not allow for additional insights.

4.2.2   Temperature

Using the available data (i.e. maximum air temperature [Tair_max], minimum air temperature [Tair_min], total rainfall [total rain], total precipitations [Total precipitation], snow accumulation [snow depth]) Royer et al. (2013) identified a tendency towards an increased mean annual temperature. This change however was not statistically significant based on the Mann-Kendall test and Theil-Sen approach. Meteorological data indicate a non-significant increase of the mean summer temperature and exception made of Chapais 2 also during the autumn season. Additionally, all three stations recorded an increase in total annual rainfall; however this was only statistically significant for the stations of Chapais 2 and Kuujjuarapik A. These two stations also had statistically significant increases for the autumn and winter season mean total rainfall. Lastly only Chapais 2 registered a statistically significant increase of the mean annual total precipitation and the winter mean total precipitation.

Similar research, such as the one by Allard et al. 2007a undertaken for OURANOS in Nunavik (and included Kuujjuarapik – the Inuit village linked with the Cree village of Whapmagoostui) using data from 1950 to 2006, showed an increase in temperature in the northern part of the James Bay territory starting in 1995. They show that 2006 was a particularly warm year with temperatures 3–4 °C higher than those recorded in 1990 (Allard et al. 2007a). These results corroborate information obtained during a similar analysis from the James Bay Advisory Committee on the Environment and previsions by Desjarlais et al. for the Ouranos Consortium using climate data for 1960–2005 (Meunier 2007; Desjarlais et al. 2010).

The majority of Cree hunters who answered the questionnaires stated that there had been an increase in temperature. This corresponded to results obtained by the CTA during their community workshops in Mistissini, Waskaganish and Whapmagoostui. The workshop participants mentioned that the temperature had increased and that the Sun's intensity was stronger. During the interviews a Mistissini participant pointed out that on average the annual temperature was higher. These observations support the recorded meteorological data.

When the participants were asked to separate their observations on the weather based on seasons, a majority of questionnaire respondents identified an increase in winter temperature. However answers for summer temperature showed less cohesion. A number of coastal participants declared not having noticed any changes in summer temperature or having noticed a cooling. During the interviews, participants mentioned warmer winters however none brought up summer temperatures. A Mistissini participant also explained that the winter seems to arrive earlier each year. This was similar to results in the community workshops where discussions on the changing climate were centred on winter conditions.

When looking at these results, it is important to keep in mind that this study and the CTA's community workshops targeted Cree hunters. As such, summer is usually a calmer period for subsistence activities with less rigorous hunting and trapping activities taking place. Hunting and trapping is most intense starting with the cold season (autumn) and ending in the spring. It is therefore conceivable that Cree hunters are more sensible to winter phenomenon as the success of subsistence activities and their safety relies more heavily on their observations. The James Bay can also influence the local climate by limiting temperature extremes and thus explaining the discrepancy with the observations on summer temperature by coastal participants. As previously mentioned, the limited number of meteorological stations, available to identify long term trends in temperature, might play an additional role in the lack of differences between the interior and costal trends (Meunier 2007; Royer et al. 2013).

4.2.3 Snowfall and Snow Depth

Questions on snowfall and depth also presented a multiplicity of answers. With participants describing that the last snowfall occurs later (40 %), earlier (33 %) and at the same time as before (20 %). The Waskaganish community workshop also discussed snowfall. The workshop participants mentioned that there were no more spring storms, that the snow was later in the fall and that it melted away sooner in the spring (CTA 2010b). However the meteorological data places the last snowfall in June in the northern part of the territory (4.7 cm at Kuujjuarapik A) and in May for the southern part (5.6 cm at Chapais 2) based on climate normal between 1971 and 2000 (Meunier 2007; Environment Canada 2011a). The data for 2000, 2001 and 2002 also does not deviate from this normal (Environment Canada 2011b). At first glance the divergence in these results appears to be related to individual perception as there is no trend base on location or age group. However it should be noted that although there continues to be snowfalls later in the spring it is accompanied by the earlier arrival of rainfall. An increase in precipitations in liquid form could strongly influence the participants' perception on whether the month is a snowy or rainy one. Additional rainfall could also lead to a faster melting of ground accumulation snow.

There is more cohesion in the responses associated with the timing of the first snowfall in autumn. A majority of participants (73 %) said that it occurs later in the year. This concurs with results obtained during the CTA community workshops where participants identified January as winter's start date and mentioned that winter was today shortened by 2 months. The responses vary based on location, with all seven participants who noted that the first snowfall occurs earlier living in coastal communities. The meteorological data does not directly support these observations as it records snowfall in September of 1.5–3.4 cm. When looked at further however, September also presents important quantities of liquid precipitations of 99–123.4 mm for the climate normals between 1971 and 2000 and for 2000, 2001 and 2002 (Environment Canada 2011a, b). This combination of solid and liquid precipitation occurs until December; with a tendency towards an increase of liquid precipitation during autumn and winter from 1975 to 2000 (Environment Canada 2011b; Royer et al. 2013). This reinforces the hypothesis that the Cree participants' observations on a later snowfall are linked to the combination of rain and snow precipitations with increasing quantities of rain. The lack of ground snow accumulation and the fact that snowfalls are often followed by large quantities of rain would make it harder for participants to remember the occurrence of the first snowfall; especially as it would not be linked to changes with the practice of traditional subsistence activities. The date given by participants would then refer to the period when there is an absence of rain. This explanation also applies for the spring season as the meteorological data for 1971–2000 show combined snow and rain precipitations over the entire territory starting in March and lasting until the end of May (Environment Canada 2011a). It is consequently not so much the moment of the first and last snowfalls that varies but rather the ratio between solid and liquid precipitations over key periods. As the participants were asked to remember far past events, what occurs is rather a general evaluation of the climate. A month that was usually more marked by snow in their youth but that today has many days of rain can then be perceived as a month without snow. Although the answers do not allow us to precisely identify the moment where the first and last snowfall occur, they do allow us to underscore the importance of an increase in rainfall on the Cree way of life and their subsistence activities.

This hypothesis is further strengthened when looking at the participants' perceptions on ground snow accumulation. A majority of questionnaire participants and two interview participants from Mistissini noted that there is a decrease in ground accumulation of snow. Additionally this decrease in quantity was also mentioned during the community workshops. Aside from mentioning that there was a decrease in ground accumulation, the community members mentioned that snow melts more quickly because the ground is not frozen and that the snow is combined with rain to create slush (CTA 2010a, b). This information is confirmed by climate studies in the region which note an increase in the permafrost's mean temperature (Desjarlais et al. 2010; Allard et al. 2007b; Thibault and Payette 2009). The James Bay territory is in a zone of discontinuous permafrost in the north and sporadic permafrost in the south (Allard and Seguin 1987). The CTA's community workshops underlined many worries linked to permafrost thaw such as infrastructure deterioration and an increased risk linked to travel because of ground instability. "Areas where permafrost is melting can be very dangerous for the hunters, they can fall" (CTA 2010c). Thus these questions on snow bring forward an important point. Although snowfall appears at the same time each year, it is its role in traditional activities that will influence the perception of change as ground accumulation of snow is necessary to allow a start of traditional winter subsistence activities. A similar observation was noted by Gearheard et al. (2010) who was working with the Inuit populations of Clyde River on wind data.

> Some additional attention to exactly what Inuit are concerned about in terms of various wind events and whether these events are likely to be captured in the weather station records are worth examining. Even a perceived change in wind direction may reflect secondary phenomena such as ice movements or snow drifts, which in turn may respond to other influences (currents, snow quality, previous storms, etc.). (Gearheard et al. 2010, p. 290)

In the case of snowfall, we should therefore not see the discrepancy as erroneous but rather an alternative understanding of the phenomena as environmental observations can rarely allow for the isolation of each phenomenon independently from its context.

4.2.4 Inland Ice Conditions

Questionnaire participants were asked to describe inland ice conditions. They used a number of visual observations to identify key changes in inland ice conditions including thickness (in inches or cm), colour (i.e. black or white), the presence of water pools and/or water pockets, and the presence of ice shards or ice candle (columnar hexagonal ice crystal).

Answers to these questions were more homogenous than those for temperature and snow. The majority of participants reported having noticed a delay in the timing of freezing of lakes and river and acceleration in the timing of ice break-up in the spring more precisely placing it in mid-April. This corresponds to the results from the CTA's community workshops. A majority of questionnaire participants also noted that the ice on lakes and rivers was thinner. This was brought up during the interviews by a number of participants who mentioned a thinner and thus more fragile ice on lakes and rivers identified through factors such as an increase in water pockets in the ice. Some participants also stated that they had noticed an absence of ice candles or ice shards in the ice which amplified its fragility and increased its melt rate. Finally other interview participants declared that the ice melted earlier in the spring. The community workshop participants reported similar observations, noting that the ice was thicker in the past and that differences in its structure were causing it to be more fragile. Conditions identified by Cree participants to be associated with weaker ice where white ice sometimes called 'soft ice' (referring to snow ice), the absence of ice shards and the presence of water pools and pockets. An ice associated with a stronger load bearing capacity is identified as being black (congelation ice) with ice shards and a distinct lack of water pools or water pockets. During the interviews, participants' comments included "The ice shards are gone and there are more water pockets and pools in the ice." (Anonymous 2010a) and "The ice isn't as strong as before, there are no more icicles in the ice." (Anonymous 2010b)

The strength of the ice is important to Cree hunters as they use lakes and rivers to move about during the hunting season (Downing and Cuerrier 2011). The decrease in its load bearing strength are a risk to both young and experienced hunters as quoted from one of the participants in the Whapmagoostui CTA workshops (2010c, p. 2): "In spring we lost two very experienced hunters that fell through the ice." The presence of increased quantity of white ice as well as its early thaw increases the risk of accidents (CBHSSJB 2010; McDonagh 2011). Similar accidents have been identified in Inuit communities in Canada (Nelson 2003; Furgal and Seguin 2006). It should be noted that the increased use of snowmobiles during the same time period have allowed hunters to go further faster on the ice increasing the severity of accidents. Travelling at high speeds also makes it harder to note precise observations on the quality of the ice. Researche done with aboriginal and Inuit groups in Canada have shown that the use of technology (e.g. GPS, VHF radios, satellite phones) procure an increased sentiment of security which combined with a decrease in the time devoted to hunting has created a situation where community members take a greater proportion of risks when they do head out hunting (e.g. going out when meteorological conditions are not optimal) (Aporta and Higgs 2005; Ford et al. 2008).

As mentioned earlier changes in mean air temperature from 1970 to 2000 were not statistically significant. Therefore the possibility that any change in ice thickness or strength is related simply to air temperature is doubtful. By combining the observations from Cree hunters (i.e. increased fragility, lack of vertical structures, increased water pockets) with the long term data from the meteorological stations (i.e. an increase in the average total precipitations in autumn and winter), it is possible to hypothesise that the observed fragility of the ice is occurring due to a change in its base structure caused by a greater rainfall. Thus a greater proportion of ice on lakes and river would be snow ice, which is formed when water saturated snow freezes on the surface contrary to congelation ice which forms columnar hexagonal ice crystal (Jones 1969; Adams and Roulet 1980; Dyck 2007; Furgal and Prowse 2008). An increased cover over lake and river surfaces would also insulate the underlying ice, decreasing heat loss and congelation ice growth. A lack of available data on historical ice composition did unfortunately not allow an explicit validation of the hypothesis.

A number of studies have looked at combining TEK and science to identify climate change's effects on the Arctic's sea ice however very few of them have looked at ice conditions on continental waters in the subarctic (Huntington 2000a; Laidler et al. 2010). It is therefore crucial to undertake further studies on the effects of climate change in the environment and more specifically on the changes in the structure and composition of ice on the Eastern James Bay territory. Such research would allow a deeper and more complete understanding of the complexities of continental ice and facilitate the development of suitable adaptation measures (Nichols et al. 2004; Laidler et al. 2010).

4.2.5   Extreme Weather Events

Interviewed hunters also mentioned that it was getting harder and harder for them to forecast the weather. They particularly mentioned having a harder time identifying incoming storms. One participant mentioned having recently been surprised by thunderstorms while hunting even though based on his normal meteorological markers he expected a calm day. In such situations hunters have only a short amount of time to take cover from the incoming weather, increasing the risk factors associated with hunting. Similar observations were also brought up by the hunters of Whapmagoostui during the CTA workshops, with participants mentioning that they could not rely on some of their traditional weather markers, saying "It is more difficult to predict the weather now" (CTA 2010c, p. 2). Studies on climate variability undertaken with the participation of Inuit populations in Canada have identified a similar increase in the difficulty of forecasting meteorological phenomena (Riewe and Oakes 2006; Nickels et al. 2006; Pearce et al. 2009). Elders and hunters in the Inuit communities mentioned not being able to base themselves on current TEK to predict meteorological events not because of any erosion in TEK knowledge but rather because of the changes in meteorological patterns (Weatherhead et al. 2010; Dixon 2010). This variability in meteorological patterns and an increase in extreme events (e.g. storms) have been proposed by many authors as being some of the first manifestations of climatic change (Huntington 2000a; Jolly et al. 2002; Fox 2002; Duerden 2004).

4.2.6   And What of Humans?

Cree hunters have observed the effects of changes in the climate on the James Bay territory. These include an increase in temperature, an increase in liquid precipitations, a decrease in the presence of permafrost, an increased fragility of the ice on lakes and rivers and a modification in traditional meteorological markers. These observations obtained through interviews and questionnaires as well as the results of the CTA's community workshops were confirmed with climate data obtained from meteorological stations. These changes imply important consequences including increased risks for hunters who now face an environment that reacts differently. They must therefore relearn or adapt their meteorological markers to this changing reality. This need for adaptation strategies and to develop tools to mitigate the effects of climate change exist both for movement on the land as for other activities linked to hunting (e.g. temporary meat conservation). Climatological previsions postulate that the effects of climate change will continue at an accelerated rate for the Canadian subarctic. Pluridisciplinary research combining TEK and scientific knowledge in the subarctic would allow for a better understanding of the effects of these changes on the local ecosystems and populations. Research associations of researchers and aboriginal community members already exist in the Arctic with the Inuit populations (e.g. ArcticNet, Arctic climate Impact Assessment) and have been greatly beneficial to knowledge development in the regions and the creation of mitigation and adaptation tools (Pearce et al. 2009; Berkes et al. 2001; Ford 2009; Berkes 2009; Gearheard et al. 2011). They identified direct and indirect effects of climate change on the health of Canadian Inuit populations such as a decrease in food security, an increase psychological distress, and the emergence of new or chronic diseases (Furgal and Seguin 2006; Furgal 2008; Parkinson and Evengård 2009; Ford et al. 2010). The adaptation strategies envisioned by these projects included community observations of climatic and environmental conditions, a modification in the hunting period timing, a change in the hunted species, waterfront protection measures, abandonment of some lands and movement of habitations (Ford and Smit 2004; Huntington et al. 2004; Berkes et al. 2007).

There is a variation in the observations based on the localisation of the communities for most climatic conditions studies. The whole of the Cree communities approached during this study noted changes (e.g. increased ice fragility, warming of the winter temperature). However the relative importance of the observed changes varies based on the communities. Those with traplines in the northern most areas of Eeyou Istchee indicated an increased preoccupation towards problems linked to permafrost thaw, while those in coastal areas were the only ones that indicated having noticed a cooling or no change in summer temperatures. They were also the only ones to note that the winter's first snow arrived earlier or at the same time as before and that there was no difference in the timing of the thaw and in the thickness of the ice on lakes and river. Additional more extensive research combining TEK and science would allow a better understanding of these variations in local observations as experienced by the local communities.

References

Adams WP, Roulet NT (1980) Illustration of the roles of snow in the evolution of the winter cover of a lake. Arctic 33(1):100–116

Allard M, Séguin MK (1987) Le pergélisol au Québec nordique: bilan et perspectives. Géog Phys Quatern 41(1):141–152

Allard M, Camels F, Fortier D, Laurent C, L'Hérault E, Vinet F (2007a) Cartographie des conditions de pergélisol dans les communautés du Nunavik en vue de l'adaptation au réchauffement climatique. Centre d'études nordiques, Université Laval, rapport soumis à Ouranos et Ressources Naturelles Canada, Québec

Allard M, Fortier R, Sarazin D, Camels F, Fortier D, Chaumaunt D, Savard J-P, Tarussov A (2007b) L'impact du réchauffement climatique sur les aéroports du Nunavik: caractéristiques du pergélisol et caractérisation des processus de dégradation des pistes. Centre d'études nordiques, Université Laval, rapport soumis à Ouranos, Ressources Naturelles Canada et Transports Québec, Québec

Anonymous (2010a) In-depth interview in Mistissini, Quebec. Interviewed by: Marie-Jeanne Royer, August 2010

Anonymous (2010b) Short interview in Mistissini, Québec. Interviewed by: Marie-Jeanne Royer, August 2010

Anonymous (2010c) Personal communications, August 2010

Aporta C, Higgs E (2005) Satellite culture: global positioning systems, Inuit wayfinding, and the need for a new account of technology. Curr Anthropol 46(5):729–753

Bahuchet S (1991) Encyclopédie des Pygmées Aka., Volume 1. Éditions Peeters, Société des Études Linguistiques et Anthropologiques de France, France

Baxter P, Jack S (2008) Qualitative case study methodology: study design and implementation for novice researchers. The Qual Rep 13(4):544–559

Berkes F (2009) Indigenous ways of knowing and the study of environmental change. J Roy Soc NZ 39(4):151–156

Berkes F, Mathias J, Kislalioglu M, Fast H (2001) The Canadian Arctic and the 'Oceans Act': the development of participatory environmental research and management. Ocean Coast Manage 44:451–469

Berkes F, Berkes MK, Fast M (2007) Collaborative integrated management in Canada's North: the role of local and traditional knowledge and community-based monitoring. Coast Manage 35:143–162

Biernacki P, Waldorf D (1981) Snowball sampling: problems and techniques of chain referral sampling. Sociol Methods Res 10:141–163

Brown RD (2010) Analysis of snow cover variability and change in Québec 1948–2005. Hydrol Process 24:1929–1954

Brunois F (2005) Pour une approche interactive des savoirs locaux: l'ethno-écologie. Le Journal de la Société des Océanistes 120–121:31–40. http://jso.revues.org/index335.html

CBHSSJB – Cree Board of Health and Social Services of James Bay (2010) Drowning in Eeyou Istchee: an overview of the death and hospitalization statistics, 1985–2007. www.creehealth.org

Conklin HC (1954) An ethnoeocological approach to shifting agriculture. Trans N Y Acad Sci 2(17):133–142

CTA – Cree Trappers' Association (2010a) The climate change project: impacts and adaptation measures for the hunters, trappers and communities of Eeyou Istchee. Mistissini Community Report – Apr 2010. Unpublished, Eastmain, Canada

CTA – Cree Trappers' Association (2010b) The climate change project: impacts and adaptation measures for the hunters, trappers and communities of Eeyou Istchee. Waskaganish Community Report – Apr 2010. Unpublished, Eastmain, Canada

CTA – Cree Trappers' Association (2010c) The climate change project: impacts and adaptation measures for the hunters, trappers and communities of Eeyou Istchee. Whapmagoostui Community Report – Apr 2010. Unpublished, Eastmain, Canada

CTA – Cree Trappers's Association (2010d) Activity report 2009–2010. Unpublished, Eastmain Canada

Cronon W (ed) (1996) Uncommon ground: rethinking the human place in nature. W.W. Norton & Company, New York

Czernochowski D, Mecklinger A, Johansson M, Brinkmann M (2005) Age-related differences in familiarity and recollection: ERP evidence from a recognition memory study in children and young adults. Cogn Affect Behav Neurosci 5(4):417–433

Desjarlais C, Allard M, Bélanger D, Blondot A, Bourque A, Chaumont D, Gosselin P, Houle D, Larrivée C, Lease N, Pham AT, Roy R, Savard J-P, Turcotte R, Villeneuve C (2010) OURANOS. Savoir s'adapter aux changements climatiques. OURANOS, Montréal

Dixon G (2010) New documentary recounts bizarre climate changes seen by Inuit Elders. Globe and Mail artid: 1763952

Downing A, Cuerrier A (2011) A synthesis of the impacts of climate change on the first nations and inuit of Canada. Indian J Tradit Knowl 10(1):57–70

Duerden F (2004) Translating climate change impacts at the community level. Arctic 57(2):204–212

Dyck MG (2007) Community monitoring and environmental change: college-based limnological studies at crazy lake (Tasirluk, Nunavut). Arctic 60(1):55–61

Ellen R (ed) (2006) Ethnobiology and the science of humankind. Blackwell, Oxford

Environnement Canada (2011a-09-14) Normales et moyennes climatiques au Canada 1971–2000. Canada. www.climate.weatheroffice.gc.ca/climate_normals/index_f.html

Environnement Canada (2011b) Rapport de données mensuelles. Canada. www.climate.weatheroffice.gc.ca/climateData/canada_f.html

Flick U (2002) An introduction to qualitative research, 2nd edn. Sage, Thousand Oaks

Ford J (2009) Dangerous climate change and the importance of adaptation for the arctic's inuit population. Environmental Research Letters 4. http://iopscience.iop.org/1748-9326/4/2/024006

Ford JD, Smit B (2004) A framework for assessing the vulnerability of communities in the Canadian Arctic to risks associated with climate change. Arctic 57(4):389–400

Ford JD, Smit B, Wandel J, Allurut M, Shappa K, Ittusarjuat H, Wrunnut K (2008) Climate change in the Arctic: current and future vulnerability in two Inuit communities in Canada. The Geogr J 174(1):45–62

Ford JD, Berrang-Ford L, King M, Furgal C (2010) Vulnerability of aboriginal health systems in Canada to climate change. Glob Environ Chang 20:668–680

Fox S (2002) These are things that are really happening: Inuit perspectives on the evidence and impacts of climate change in Nunavut. In: Igor K, Dyanna J (eds) The Earth is faster now: indigenous observations of Arctic environmental change. Arctic Research Consortium of the United States, Fairbanks

Furgal C (2008) Health impacts of climate change in Canada's North. In: Séguin J (ed) Human health in a changing climate: a Canadian assessment of vulnerabilities and adaptive capacity. Health Canada, Ottawa

Furgal C, Prowse TD (2008) Northern Canada. In: Lemmen DS, Warren FJ, Lacroix J, Bush E (eds) From impacts to adaptation: Canada in a changing climate 2007. Government of Canada, Ottawa

Furgal C, Seguin J (2006) Climate change, health, and vulnerability in Canadian northern aboriginal communities. Environ Health Perspect 114(12):1964–1970

Gearheard S, Pocernich M, Stewart R, Sanguya J, Huntington HP (2010) Linking Inuit knowledge and meteorological station observations to understand changing wind patterns at Clyde River, Nunavut. Clim Change 100:267–294

Gearheard S, Aporta C, Aipellee G, O'Keefe K (2011) The Igliniit project: inuit hunters document life on the trail to map and monitor Arctic change. The Can Geogr 55(1):42–55

Geertz C (1973) Thick description: towards an interpretive theory of culture. In: Geertz C (ed) The interpretation of cultures: selected essays. Basic Books, New York

Ghetti S, Angellini L (2008) The development of recollection and familiarity in childhood and adolescence: evidence from the dual-process signal detection model. Child Dev 79(2):339–358

Godelier M (1984) L'idéel et le matériel. Fayard, Paris

Gumuchian H, Marois C (2000) Initiation à la recherche en géographie. Les Presses de l'Université de Montréal, Montréal

Haraway D (1988) Situated knowledges: the science question in feminism and the privilege of partial perspective. Fem Stud 14(3):575–599

Huntington HP (2000a) Native observations capture impacts of sea ice changes. Witn Arctic 8 (1):1–2

Huntington HP (2000b) Using traditional ecological knowledge in science: methods and applications. Ecol Appl 10(5):1270–1274

Huntington HP, Suydam RS, Rosenberg DH (2004) Traditional knowledge and satellite tracking as complementary approaches to ecological understanding. Environ Conserv 31(3):177–180

Johnson LM (2010) Trail of story, traveller's path: reflections on ethnoecology and landscape. AU Press, Edmonton

Jolly D, Berkes F, Castleden J, Nichols T (2002) We can't predict the weather like we used to: Inuvialuit observations of climate change, Sachs Harbour, western Canadian Arctic. In: Krupnik I, Jolly D (eds) The earth is faster now: indigenous observations of Arctic environmental change. Arctic Research Consortium of the United States, Fairbanks

Jones JAA (1969) The growth and significance of white ice at Knob Lake, Quebec. Can Geogr 13(4):354–372

Jones PD, Osborn TJ, Briffa KR (1997) Estimating sampling errors in large-scale temperature averages. J Climate 10:2548–2568

Laidler GJ, Elee P, Ikummaq T, Joamie E, Aporta C (2010) Mapping sea-ice knowledge, use and change in Nunavut, Canada (Cape Dorset, Igloolik, Pangnirtung). In: Krupnik I, Aporta C, Gearheard S, Laidler GJ, Kelsen-Holm L (eds) SIKU: knowing our ice, documenting inuit sea-ice knowledge and use. Springer, Dordrecht

MacCormack C, Strathern M (eds) (1980) Nature culture and gender. Cambridge University Press, Cambridge

Maffi L (2005) Linguistic, cultural, and biological diversity. Annu Rev Anthropol 34:599–617

McDonagh P (2011) Observing the effects of climate change in Eeyou Istchee. Public Health Department of the Cree Health Board, Quebec

Meunier C (2007) Portrait and known environmental impacts of climate change on the James Bay territory. James Bay Advisory Committee on the Environment, Quebec

Milton K (1997) Ecologies: anthropologie, culture et environnement. Rev Int Sci Sociales 154:519–538

Mülhäusler P (1995) The interdependence of linguistic and biological diversity. In: Myers D (ed) The politics of multiculturalism in the Asia/Pacific. North Territory University Press, Darwin

Nelson O (2003) Two men drown in Inukjuak: snowmobile crashes through thin ice. Nunatsiaq News

Nichols T, Berkes F, Jolly D, Snow NB (2004) The community of Sachs Harbour, climate change and sea ice: local observations from the Canadian Western Arctic. Arctic 57(1):68–79

Nickels S, Furgal C, Buell M, Moquin H (2006) Unikkaaqatigiit, putting the human face on climate change, perspectives from inuit in Canada. Inuit Tapiriit Kanatami, Nasivvik Centre for Inuit Health and Changing Environments at Université Laval and the Ajunnginiq Center at the National Aboriginal Health Organization, Ottawa

Paillé P (1996) De l'analyse qualitative en general et de l'analyse thématique en particulier. Rech Quant 15:179–194

Parkinson AJ, Evengård B (2009) Climate change, its impact on human health in the Arctic and the public health response to threats of emerging infectious diseases. Clim Change Infect Dis. Glob Health Act 2. doi:10.3402/gha.v210.2075

Pearce TD, Ford JD, Laidler GJ, Smit B, Duerden F, Allarut M, Andrachuk M, Dialla A, Elee P, Goose A, Ikummaq T, Joamie E, Kataoyak F, Loring E, Meakin S, Nickels S, Shappa K, Shirley J, Wandel J (2009) Community collaboration and climate change research in the Canadian Arctic. Polar Res 28:10–27

Riewe R, Oakes J (eds) (2006) Climate change: linking traditional and scientific knowledge. Aboriginal Issues Press, University of Manitoba, Manitoba

Royer M-JS, Herrmann TM, Oliver S, Kénel D, Daniel F, Rick C (2013) Linking Cree hunters' and scientific observations of changing inland ice and meteorological conditions in the subarctic Eastern James Bay region, Canada. Clim Change 119(3):719–732

Sandelowski M (1995) Sample size in qualitative research. Res Nurs Health 18:179–183

Small ML (2009) 'How many cases do I need?' On science and the logic of case selection in field-based research. Ethnography 1091:5–38

Stake RE (2005) Qualitative case studies. In: Denzin NK, Lincoln YS (eds) The SAGE handbook of qualitative research, 3rd edn. Sage, Thousand Oaks

Sutherland WJ (2003) Parallel extinction risk and global distribution of languages and species. Nature 423:276–279

Thibault S, Payette S (2009) Recent permafrost degradation in bogs of the James Bay area, Northern Quebec, Canada. Permafr Periglac Process 20:383–389

Toledo VM (1992) What is ethnoecology? Origins, scope and implications of a rising discipline. Etnoecológica 1(1):5–21

Toledo VM (2002) Ethnoecology: a conceptual framework for the study of indigenous knowledge of nature. In: Stepp JR, Wyndham FS, Zarger RK (eds) Ethnobiology and biocultural diversity: proceedings of the seventh International congress of ethnobiology. International Society of Ethnobiology, Athens

Watanabe H (1964) The Ainu: a study of ecology and the system of social solidarity between man and nature in relation to group structure. Journal of the Faculty of Science 2(6). University of Tokyo, Section V, Anthropology, Tokyo, Japan

Watanabe H (1983) Occupational differentiation and social stratification: the case of Northern Pacific Maritime food-gatherers. Curr Anthropol 24(2):217–219

Weatherhead E, Gearheard S, Barry RG (2010) Changes in weather persistence: insight from Inuit knowledge. Glob Environ Chang 20:523–528

Williams M (2000) Interpretivism and generalisation. Sociology 34(2):209–224

Yin RK (2003) Case study research: design and methods, 3rd edn. Sage, Thousand Oaks

Chapter 5
Conclusions

This book and the associated research combined different sources of observations (i.e. TEK and scientific) to study the changes in climatological conditions over the last 30 years in the James Bay territory. This process had two main objectives (1) identify what observations and meteorological station data could reveal about current impacts of climate change; and (2) how contrasting data sets can be combined to offer a more detailed and insightful view of the phenomena at play. Traditional ecological knowledge (TEK) is a complex and dynamic system of knowledge, representations and practices linked to nature and the environment; it and scientific knowledge should be regarded as different but complementary knowledge that can be used to enrich themselves mutually. In the subarctic this combination can be used by offering additional base data on changes to meteorological conditions, environmental and sociocultural behaviour and by helping elaborate research questions and hypothesis.

For research into the impacts of climatic changes on the subarctic a broader definition of climatic change which includes both anthropogenic and natural causes was more suitable and was used in this case. It allowed and will allow researchers to concentrate on all the impacts currently experienced by populations living in the region without getting into the debate of its root cause. Finally the need to use an interdisciplinary approach must be stressed when undertaking a case study in an area that has undergone such large scale changes to its social, environmental and physical environment.

The James Bay territory has undergone a large number of environmental changes starting in the second half of the twentieth century, most notably with the creation of the La Grande hydroelectric complex. These environmental changes coincided with what is identified by climate research as the exponential increase in current climate change. The context of intense sociocultural and economic changes in the James Bay territory touches many societal aspects including food, education, social organisation and language. The extent of the environmental and social changes would be enough to identify the James Bay territory as a vulnerable space which would then be increasingly destabilised by the impacts of climate

© The Author(s) 2016
M.-J.S. Royer, *Climate, Environment and Cree Observations*,
SpringerBriefs in Climate Studies, DOI 10.1007/978-3-319-25181-3_5

change. However in Eeyou Istchee subsistence activities continue to be practiced regularly by a large proportion of community members. The Cree have been incorporating new knowledge and technologies in their way of life while adapting them to their reality and culture.

The results of this study indicate that there are real changes in the environment and climate of the James Bay territory. Cree participants revealed concerns about changes pertaining to winter temperature, precipitations and thaw and freeze-up periods of inland ice. These concerns vary slightly based on the community to which the participants belong (e.g. participants in communities or with traplines further north are more concerned by the consequences linked to permafrost thaw). The combination of TEK and science allowed us to better orient the explanations for some phenomena. For example, after having noted that participants had noted an increase in inland ice, meteorological data was analysed to find potential causes of this fragility. In that case, the ice characteristics mentioned by participants were combined to long term trend analysis and raised the possibility of increased precipitation being the cause. This and similar results seem to point towards climate observation being influenced by changes in the climate rather than resting solely on the impacts of anthropic changes to the physical environment. Many of the changes seem to revolve around an increased quantity of liquid precipitation in the autumn and winter.

Although promising the results here are limited by the small number of meteorological stations and their geographic distribution as well as by the limited sample size of the qualitative research. Because of this, there were difficulties in distinguishing differences in temperature trends based on location or time period. These preliminary results underscore the need for supplementary research (both qualitative and quantitative) in addition to community monitoring programs. The increased information on changes in the conditions in the James Bay territory would allow the development of targeted adaptation strategies, and more exact analysis of changes to the conditions of inland ice, permafrost and weather conditions. Potential adaptation strategies could include awareness programs, and the creation of community-based updates on winter conditions (e.g. ice thaw and breakup) and extreme weather sightings. Future measurement initiatives should include and emphasise local community indicators (e.g. ice colour, ice shards, proportion of liquid precipitations in the winter) which would assure a better correspondence between scientific and local observations and help translate the results into meaningful information for the concerned communities.

Changes in the environment have increased risks associated with travel and the practice of traditional subsistence activities. This in turn could decrease access to traditional food resources which could cause a decrease in food security (Royer and Herrmann 2011). With traditional foods essential to Cree well-being a decrease in its availability could negatively impact the local populations. "I remember in Old Nemaska some people receive a few sickness and they want to eat real food. When they eat real food, in 2 days the pain is gone (sic)" (Anonymous 2010). Research in the Arctic with Inuit populations have shown that a decreased practice of traditional subsistence activities is associated with increased psychological distress, decreased food security, and the emergence of chronic and/or new diseases (Furgal and Seguin 2006; Parkinson and Evengard 2009; Ford et al. 2010).

Current research in the territory is fragmented, concentrating on one particular aspect or village. Although pertinent and useful they should be complemented with research that observes the region as a whole. Such research would allow the identification of larger scale tendencies for the area. The James Bay territory, with its massive land surface area, is part of a transition zone which would benefit greatly from north-south and east-west trend analysis of the impacts of climate change. As such, this book and the research behind it represents only a small part of what is an interesting and diverse field of study.

References

Anonymous (2010) Personal communications, August 2010

Ford JD, Berrang-Ford L, King M, Furgal C (2010) Vulnerability of aboriginal health systems in Canada to climate change. Glob Environ Chang 20:668–680

Furgal C, Seguin J (2006) Climate change, health and vulnerability in Canadian northern aboriginal communities. Environ Health Perspect 114(1):1964–1970

Parkinson AJ, Evengard B (2009) Climate change, its impact on human health in the Arctic and the public health response to threats of emerging infectious disease. Climate change and Infectious Diseases. Global Health Action 2. doi:10.3402/gha.v210.2075

Royer M-JS, Herrmann TM (2011) Socio-environmental changes in two traditional food species in the Cree First Nation of subarctic James Bay. Cah Géogr Québec 55(156):575–601

Glossary

Black ice Ice formed by congelation whose crystals are oriented in a regular fashion creating vertical structures. As it is transparent, it allows the colour of the water beneath it to show through making it look black. It does not include any air bubbles

Chishaaminituu Great spirit who offered Eeyou Istchee to the Cree

Eeyou Name used by the Cree to identify themselves

Eeyou Istchee Name given by the Cree to their ancestral lands

Eeyou Weeshou-Wehwun Name of the Cree's traditional laws. They are usually transmitted in oral form

Glacial isostatic adjustment It consists of the rise of land masses that were previously depressed by ice sheets

Ice candle Columnar hexagonal ice crystals formed in congelation ice (black ice). They are especially visible on water masses during the thaw. Also called ice shard

Ice shard Columnar hexagonal ice crystals formed in congelation ice (black ice). They are especially visible on water masses during the thaw. Also called ice candle

Indoh-hoh Istchee Ouje-Maaoo Cree name for the Tallyman

Kaanoowapmaakin Cree name for the Tallyman

Ojibwe Indigenous First Nation group belonging to the Algonquien language family group

Tallyman Person responsible for the management and conservation of a family trapline and its history

Trapline Familial hunting, trapping and gathering grounds. Called Indoh-hoh Istchee in Cree

White ice Ice formed when the snow accumulated at the surface of water is imbibed with water and then frozen. Its structure includes air bubbles and is opaque (white) with crystals oriented haphazardly

© The Author(s) 2016
M.-J.S. Royer, *Climate, Environment and Cree Observations*,
SpringerBriefs in Climate Studies, DOI 10.1007/978-3-319-25181-3

Printed in the United States
By Bookmasters